Contemporary Discourse
in the Field of
ASTRONOMY ™

The Role of the Sun in Our Solar System

An Anthology of Current Thought

Edited by Jennifer Viegas

The Rosen Publishing Group, Inc., New York

Published in 2006 by The Rosen Publishing Group, Inc.
29 East 21st Street, New York, NY 10010

First Edition

Library of Congress Cataloging-in-Publication Data

The role of the sun in our solar system: an anthology of current
thought/edited by Jennifer Viegas.—1st ed.
 p. cm.—(Contemporary discourse in the field of astronomy)
Includes bibliographical references.
ISBN 1-4042-0396-6 (library binding)
1. Sun—Popular works. 2. Solar system—Popular works.
I. Viegas, Jennifer. II. Series.
QB521.4.R65 2006
523.7—dc22
                                                    2004027661

Manufactured in the United States of America

On the cover: Bottom right: an extreme ultraviolet imaging
telescope (EIT) image of a huge handle-shaped solar prominence,
taken on September 14, 1999. Bottom left: Galileo Galilei.
Center left: the Dumbell Nebula. Top right: solar flares.

CONTENTS

Introduction

A star was born 4.6 billion years ago. While this star is just one of 100 billion other celestial bodies in the Milky Way galaxy, it shines brighter than any other star ever observed. Both figuratively and literally, this gleaming orb that we call the Sun is the true shining star of our solar system.

Here on Earth, all life relies upon energy that the Sun provides. Without the Sun, Earth would be a dark, lifeless place frozen in time. But it is impossible to separate Earth from the Sun. The two have been intertwined since the beginning of our solar system.

While no one can say with certainty what happened more than 4 billion years ago, current theory holds that an immense cloud of dust and gas began to slowly spin. Gravity pulled together a dense region within this cloud. The pull caused the speed of the spin to increase, similar to how an ice skater's spin picks up speed when weight is drawn into the center of the spin. The spinning caused the gas in the center to heat up, which led to reactions that converted the dust and gas into solid matter. The resulting matter collected around the hot center as planetesimals. These planetesimals then joined in areas to form the planets. Over time, the central region became so hot and dense that nuclear fusion began to occur. This celestial version of a natural, pollution-free nuclear reactor evolved into the Sun.

The Sun's Composition

We can see the Sun's light and feel its warmth, but most of us are not conscious of the wash of Sun particles that continuously flow over our bodies. Earth is bathed in these particles, which can be collected and measured, re-created in laboratory settings, or studied using known principles of astrophysics and chemistry. Because of such research, astronomers have a good idea of what makes up the Sun.

Most of the Sun consists of hydrogen, an element that is common to Earth and all of the other planets. Additional elements exist on the Sun, but in much smaller amounts. Every 1 million atoms of hydrogen on the Sun corresponds to 98,000 atoms of helium, 850 atoms of oxygen, 360 atoms of carbon, 120 atoms of neon, 110 atoms of nitrogen, 40 atoms of magnesium, 35 atoms of iron, and 35 atoms of silicon. The names of these elements likely are familiar to you, but the form they take on the Sun is an unusual type of gas called plasma, which is sensitive to magnetism.

The entire Sun is a giant mass of plasma that is divided into distinct layers. Astronomy experts refer to these layers as zones. At the core of the Sun exists all of the pressure that supports the rest of the star. The core is also the site for all of the Sun's nuclear fusion. During this process, heat and pressure cause single protons of hydrogen atoms to join together. The atomic activity produces a hydrogen atom with a nucleus, four particles, and a separate, neutral particle called a neutrino. The mass of the proton-rich structure is less than the sum of its parts, and the loss of mass converts into energy. It is this energy that radiates throughout the solar system, including to Earth, through the Sun's zones. They are, in order from their proximity to the core, the radiative zone, the convection zone, the photosphere, the chromosphere, the transition region, and the corona. When you look up at the Sun on a sunny day, you actually see the photosphere, which emits all of the Sun's visible light.

A Dynamic, Ever-Changing Star

As if continual nuclear reactions were not enough, the processes of the Sun are subject to changes resulting from the rotation of the glowing ball around its axis and the impact of electromagnetic forces generated by the nuclear fusion, heat, and pressure. These changes can lead to eruptions of magnetic fields that tear through the Sun's zones. The fields, in turn, can result in solar activity that includes visible events such as sunspots, flares, and coronal mass ejections. You might have even seen a sunspot, which looks like a dark patch on the surface of the Sun. While it is not wise to directly gaze at the Sun, sunspots can be seen with the naked eye. They are even more visible when seen through a telescope with special filters.

The Sun has its own unique form of weather that follows measurable cycles and yet is also ever-changing, not unlike the often unpredictable weather forces here on Earth. Solar storms rain super-charged electromagnetic particles down to Earth. This storm activity can affect satellite systems and any other kind of equipment that involves magnetism or electricity. Because we rely upon computers, satellites, and so many other electricity-dependent items, concern over solar storm activity has increased over the years.

Concern has also heightened over how much the Sun warms Earth, since the light and heat emitted by the Sun varies depending on points within the Sun's cycle and solar weather events. Activities here on Earth, such as possible changes to the atmosphere and to our protective ozone layer by pollution, might also affect how the Sun governs our climate and weather. Climate change has been blamed for adverse health effects and even species extinctions, but the issue remains as hot as the Sun is itself. What most experts do agree upon is that the fate of Earth, and all of its living creatures, is intertwined with the Sun and solar activity.

A New Genesis

The word "genesis" refers to origin, and, in a way, it can relate to where science today stands via the Sun. Technology has put us at the very tip of solar exploration. This became most evident with the crash landing of the *Genesis* spacecraft on September 8, 2004. The mission of *Genesis* was to collect samples from solar wind, the whirl of charged particles that flows around and out of the Sun. As of this writing, the fate of those samples, stored in approximately 3,000 containers, remains unclear. A relatively minor glitch, a failed parachute, led to the spacecraft's hard landing.

Although the *Genesis* mission was not perfect, and no man has ever yet, or probably ever will, come close to approaching the unimaginably hot 11,000-degree Fahrenheit (6,093-degree Celsius) surface of the Sun, you will read in this anthology about the tremendous strides that have been made toward understanding the Sun and its processes. You will learn how scientists study the Sun with little direct material to work with. You will also learn about solar wind, solar storms, ultraviolet-B radiation, and Sun cycles, and how these events can impact our life on Earth. You will learn how birds, bees, and all living creatures depend upon the Sun for their survival. Finally, you will learn about some of the latest innovations concerning our use of solar energy, which could forever transform Earth into a cleaner, healthier, and more productive planet. —*JV*

How Scientists Study the Sun

The Sun is 92,960,000 miles (149,604,618 kilometers) from Earth. Its outer visible layer, the photosphere, has a temperature of around 11,000 degrees Fahrenheit, or 6,093 degrees Celsius. Both the Sun's distance from Earth and its tremendous heat prevent scientists from directly analyzing it. Instead, researchers now primarily rely upon two-dimensional images to develop theories about the three-dimensional Sun and its three-dimensional processes.

The paper "Enhanced: Imaging the Sun's Eruptions in Three Dimensions" reveals how scientists are overcoming the loss in dimension and perspective. Three new techniques are described. Polarization, or the assumption of light waves into definite forms, can be measured to estimate the third dimension. STEREO (Solar-Terrestrial Relations Observatory) imaging involves two spacecraft that take simultaneous photographs of the Sun from different angles. Finally, spectroscopic methods measure inferred distances

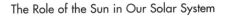

between light waves to formulate theories about solar events. Together, the three techniques enable astronomers and others to study the Sun, even though no human can set foot on the unimaginably hot surface of the distant, luminous celestial body. —JV

"Enhanced: Imaging the Sun's Eruptions in Three Dimensions"
by John C. Raymond
Science, July 2, 2004

Observations of the Sun and the solar corona reveal complex structures in exquisite detail, and they give us the opportunity to understand the processes that heat the corona and drive the solar wind. However, interpretation of the observations is often ambiguous because we observe three-dimensional structures projected onto two-dimensional images. In particular, the stressed magnetic field configurations that lead to violent solar events are inherently three-dimensional. Among the most interesting are solar flares (the huge explosions from the Sun's surface that occur over several minutes) and coronal mass ejections, or CMEs (bubbles of gas that emerge over hours; see the [first figure in the original article]). As Moran and Davila report, it is possible to recover a rich variety of three-dimensional structural information about these events (1) from two-dimensional data.

When solar activity is at a minimum, the coronal structure is stable enough that observations over a 27-day rotation period can be used for a tomographic reconstruction, much like a medical CAT (computerized axial tomography) scan. This has been done for white light scattered from electrons (2) and for individual ultraviolet spectral lines (3). Transient events such as CMEs, however, cannot be treated by tomographic methods, yet they are extremely important. Although they carry only a modest fraction of the mass in the solar wind, they can produce violent "space weather" events that can disable satellites and endanger astronauts (see the [second figure in the original article]). For example, a CME at the end of October last year caused spectacular auroral displays, but also caused problems with power grids and communications.

The three-dimensional structure of CMEs is especially important because it is related to the direction of the magnetic field, which is a controlling factor in the strength of the space weather effects. Moreover, the helical structure observed in many CMEs is related to magnetic helicity. Although there is active debate about this, conservation of magnetic helicity may govern CME evolution in interplanetary space (4), and CMEs may play an important role in the dynamo process that generates solar magnetic fields by shedding magnetic helicity (5).

One technique for inferring the three-dimensional structure is to measure Doppler shifts of spectral lines and combine the inferred velocities along the line of sight with a series of images that show the structure in the plane of the sky (6). Another technique is to image the CME from different points of view. The two STEREO (Solar-Terrestrial Relations Observatory) spacecraft scheduled for launch in November 2005 will

move apart from each other along the Earth's orbit to produce stereoscopic images.

A new technique for recovering information about the three-dimensional structure is demonstrated by Moran and Davila (1). When light is scattered by electrons in the CME it becomes polarized, and its polarization fraction depends on the angle of the scattering. Thus, the measured polarization can be used to determine the scattering angle. For each point in a coronagraph image, the method yields a weighted average distance of the scattering plasma from the plane of the sky. Given that information, Moran and Davila construct images of CMEs observed by the LASCO (Large Angle and Spectrometric Coronagraph) experiment aboard the SOHO (Solar and Heliospheric Observatory) spacecraft as they would appear from above the solar north pole or from a position to the east or west of the Sun.

The new method promises to be important for estimating the angle at which a CME emerges from the Sun, and therefore the likelihood that it will strike the Earth. It will also help resolve questions about CME structure. For example, there is still debate about whether the roughly circular structures at the leading edge of a CME should be interpreted as the projections of more or less spherical shells or as bright expanding loops. The reconstruction method can be applied to many existing CME observations obtained by LASCO.

The polarization technique, STEREO imaging, and spectroscopic techniques are all mutually complementary in that they provide different ways of viewing a solar ejection structure. The polarization technique gives an average distance from the plane of the sky, whereas spectra provide the distribution along the line of sight of a selected, denser gas at specific temperatures. STEREO images provide a full three-dimensional picture, but only the simplest structures will be amenable to a direct reconstruction, and most will require interpretation based on models. Together, these methods will greatly enhance the capability for forecasting space weather and the understanding of the physical processes that drive CMEs.

References

1. T. G. Moran, J. M. Davila, *Science* 305, 66 (2004); published online 27 May 2004 (10.1126/science.1098937).
2. R. Frazin, P. Janzen, *Astrophys. J.* 570, 408 (2002) ADS] [Abstract].
3. A. Panasyuk, *J. Geophys. Res.* 104, 9791 (1999) ADS] [AGU].
4. A. Kumar, D. M. Rust, *J. Geophys. Res.* 101, 15667 (1996) [ADS] AGU].
5. B. C. Low, *J. Geophys. Res.*, 106, 25141 (2001) ADS] [AGU].
6. A. Ciaravella *et al.*, *Astrophys. J.* 529, 575 (2000) [ADS] [Abstract/full text].

In addition to studying images of the Sun, scientists attempt to replicate the Sun's processes in laboratories right here on Earth to gain a better understanding of both basic and complex events that take place within and around the Sun. A number of those lab experiments are discussed in "The Sun Also Writhes: Laboratory Solar Physics Sheds First Light on Sol's Seething Sinews."

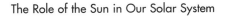

Physicist Paul Bellan, for example, electrifies hydrogen gas in his laboratory at the California Institute of Technology. About 94 percent of the Sun's atoms are hydrogen. This element is key to the production of the giant star's light and energy. When Bellan zaps hydrogen with electricity, the hydrogen gas turns into plasma, which is a type of gas that is sensitive to magnetism. Most scientists believe electricity and magnetism—the force between currents of electricity—are the two primary forces that drive the solar plasma's characteristics and movement. Bellan's work is helping to shed light on the evolution, motion, and eruption of prominences, or raised areas of plasma on the Sun. In other research, University of Wisconsin–Madison scientists have re-created a plasma flow in a test tube. Additional experiments at other universities and labs are presented in the article.—JV

"The Sun Also Writhes: Laboratory Solar Physics Sheds First Light on Sol's Seething Sinews"
by Peter Weiss
Science News, March 27, 1999

Paul Bellan creates tiny solar eruptions in his laboratory. He admits that they're not true miniatures of events on the sun. Yet they are faithful enough, he believes, to provide new clues to some important solar mysteries.

The sun is a huge ball of gases—but not electrically neutral gases as in Earth's atmosphere. Nuclear fusion at the core of the sun creates conditions of astounding heat, fast-moving particles, and mighty electric and magnetic fields. These forces shear the electrons from atoms in ordinary gases, transforming the gases into clouds of charged particles called plasmas.

Plasmas exhibit extraordinarily complex behavior not seen in other fluids. Scientists have been studying them in the laboratory for decades in pursuit of nuclear fusion as a commercial energy source. Recently, Bellan, who is at the California Institute of Technology, and a small cadre of other plasma physicists have begun to apply their laboratory methods to the study of the sun.

Many intriguing solar features arise from plasma's peculiar behavior. Tumultuous conditions sculpt the plasma at the sun's surface into huge arches of fluid flame known as solar prominences. In the intensely hot outer atmosphere, or corona, collisions of huge blobs of plasma fuel enormous jets of energy called solar flares, which often fire far into space.

In other spectacular discharges called coronal mass ejections, the corona sloughs off not just energy but also huge billows of hot plasma, sending billions of tons of the sun itself into space. These ion clouds sometimes crash into Earth with sorry consequences for satellites and power grids. One such speeding plasma bomb hammered Earth's magnetic field in March 1989, causing electromagnetic disruptions that blacked out the Canadian province of Quebec. Another January 1997 mass ejection may have fried an AT&T satellite.

These and many other features of the sun are poorly understood because solar measurements are sparse and theories of plasma physics are imperfect. "Often in solar physics, you are stuck trying to infer from an incomplete set of measurements why things occur," says Richard Canfield of Montana State University in Bozeman.

The only tools available for solar study have been ground-based telescopes and satellites. Canfield and his fellow observers "can't stick in a probe and actually make a measurement like you could in a laboratory," he says.

Skeptical at first that a device that fits in a laboratory could yield anything meaningful about a realm as vast as the sun, many solar observers are now finding themselves pleasantly surprised.

They admit that such laboratory experiments are yielding new insights into solar phenomena on multiple levels—from the looping plasmas on the sun's face to the rushing torrents of plasma believed to flow in its depths. The lab scientists are also tackling fundamental aspects of plasma behavior that may have a bearing on a range of solar features.

"I believe [the laboratory experiments are] going to lead eventually to a much clearer understanding of solar activity," says David Rust, a solar observer at Johns Hopkins University Applied Physics Laboratory in Laurel, Md.

During an experiment in Bellan's lab, the naked eye sees a single pink flash. A high-speed camera, however, shows a miniature, glowing arch expanding from the end of a horseshoe-shaped magnet. The curve builds, wavers, twists, and then dissipates into the surrounding vacuum—all in a matter of microseconds.

There's no comparison in grandeur between such pip-squeak arcs and the mighty solar prominences that Bellan is trying to replicate. The sun's contorted arches of plasma as hot as tens of thousands of degrees Celsius can tower up to 100,000 kilometers above the churning solar surface. They also can linger for days or weeks. Nonetheless, a variety of shared characteristics makes the resemblance good enough, Bellan says.

In his experiments, Bellan zaps puffs of hydrogen gas with hundreds of megawatts of electricity to transform the gas into plasma and to make the arcs twist in response to their own magnetic forces, as many of their big solar cousins do.

"We can actually duplicate something going on in the sun that people thought [was explained by] a very different physics," he says.

The simulations may shed new light on how prominences arise, contort, and erupt, Bellan explains. He and his colleague Freddy Hansen reported their work last November at the American Physical Society's Division of Plasma Physics meeting in New Orleans.

The simulations have already led them to a new, well-received model for the formation of bright, S-shaped features on the solar surface that appear to lead to solar eruptions within hours or days. In a 1996 study of some 50 such filaments seen in satellite observations, researchers led by Rust concluded that the S shapes were top-down views of extraordinarily hot, twisted prominences. Some scientists suspect that

if prominences get too twisted, they become unstable and erupt.

In a model developed by Bellan, a plasma's own magnetic forces produce the S shapes, which also appear in computer simulations of his lab arches. The magnetic forces twist the plasma until its current flows parallel to its magnetic field lines, Bellan explains.

Some researchers, such as John T. Gosling at Los Alamos (N.M.) National Laboratory, however, don't agree that the S is a contorted prominence. Data collected from erupting solar plasma ejected into space have shown that prominences within it remain at their typical temperatures, he says. The S shapes, however, appear to reach coronal temperatures of 1 to 2 million degrees. In observing the S's, "I think we're probably looking down at material surrounding the prominence, rather than the prominence itself," he comments.

Whether on the sun or in a laboratory, plasmas flow in a complicated manner that taxes the ability of scientists to model and predict their course. The complexity arises from the fundamental interdependence of electricity and magnetism. When a plasma moves, an electric current flows—by definition—because the particles of the plasma are charged. But current flow creates a magnetic field, and magnetic forces act on moving charges. So, a moving plasma pushes itself around and by doing so creates new forces that push it yet some more.

In the sun's deep interior, scientists believe, such turbulence somehow generates a huge, orderly river of current—the solar dynamo—that creates the star's magnetic field. Variability in this dynamo probably also drives the 11-year cycle of surface activity.

Researchers at the University of Wisconsin–Madison have reproduced that smooth stream of charge in a laboratory plasma. When they create a hydrogen plasma held by magnetic forces inside a doughnut-shaped vessel, tumultuous magnetic fluctuations "like splashing in a bathtub" arise spontaneously, says Wisconsin's Stewart C. Prager.

To measure the fluctuations, the Wisconsin researchers salt their plasma with carbon atoms that they use as tracers. Then, by observing shifts in the wavelength of the radiation emitted by the plasma-heated tracers, the scientists determine plasma motions. Magnetic sensors studding the outside of the vessel, which is about 5 meters in diameter, also reveal the shifting magnetic-field directions inside.

"These magnetic fluctuations, we believe, lead to the dynamo effect in our laboratory and in the sun," Prager says. Both the reasons for the turmoil and the way it produces an orderly electric flow are the quarry of the research group he leads.

For solar science, deeper understanding of plasmas is the ultimate aim. For fusion energy research, however, the scientists also hope to find a way to banish the dynamo effect from reactors because it diverts energy that could otherwise heat the plasma and sustain fusion reactions.

Even a plasma in the lab remains difficult to study. Developing the technologies for measuring their hot, highly charged, magnetized plasma has

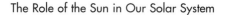
proven challenging and time-consuming. Five years ago, the researchers were able to take readings only at the fringes of the plasma. In the May PHYSICS OF PLASMAS, they will describe finally probing the 5-million-degree core.

Another fundamental solar mechanism yielding to laboratory study is magnetic reconnection, which is the breaking and reattaching of magnetic field lines that have snapped after twisting, stretching, or crossing other lines. To observe reconnection, Masaaki Yamada and his colleagues at the Princeton Plasma Physics Laboratory in New Jersey are slamming together pairs of doughnut-shaped plasma clouds known as spheromaks.

The phenomenon is attracting interest from solar and fusion researchers because it may be accompanied by powerful energy releases and therefore play a key role in features of the sun. Some scientists suggest that reconnection accounts for the mysteriously high temperature of the corona. At more than a million degrees, the corona blazes at 200 times the temperature of the solar surface. Reconnection may also energize solar flares, which explode from the corona. They typically attain temperatures in the tens of millions of degrees.

Fusion researchers—Yamada's group counts among them—also recognize magnetic reconnection as an important aspect of reactor plasmas that may be useful in boosting plasma temperatures.

Although suspecting that reconnection plays a leading role in the solar drama, theorists have struggled for decades to explain how. Plasmas, especially on the sun, are wispy gases, but the magnetic fields threading through them make them behave as if they were viscous fluids, flowing and intermingling slowly. According to the classical theory of plasmas, magnetic field lines cannot reconnect or, at best, can do so only at a stately pace because of this viscosity. This model is obviously incomplete because it would require millions of years for solar flares to release the energy they expel in minutes or hours.

To resolve the dilemma, researchers have proposed revisions to the classical picture. They postulate that the reconnections occur under extraordinary conditions of electric current flow, perhaps accompanied by shock waves, at the narrow boundaries where magnetic fields with opposite directions clash. Until Yamada's group began its experiments 3 years ago, no one had attempted to detect such conditions.

The results of the Princeton experiments don't quite match any of the theories of reconnection advanced so far, Yamada explains. As described in the April 13, 1998 PHYSICAL REVIEW LETTERS, he and his colleagues see reconnection taking place at a pace much faster than classical theory would allow but still only a hundredth of the rate required to explain solar flares. Also, within the turmoil of the spheromaks' collision, they see no evidence of shock waves.

Uncertain how to explain their findings, the Princeton researchers suggest that they may have discovered a new phenomenon that none of the previous theories included. It's a turbulence in the plasma that would increase interactions between plasma particles and thereby promote reconnections.

The researchers have recently enjoyed some confirmation of the relevance of their lab work to actual circumstances in space. They measured the thickness of the narrow boundary where two, magnetically opposite plasmas collide and cancel each other's magnetic fields. The as-yet-unpublished measurement is in keeping with satellite readings from the region of space where the magnetic fields of the sun and Earth meet, Yamada says.

As the fields of solar physics and fusion research meet and begin to blend, both disciplines are getting a shot of energy. Fusion scientists are awakening to the sun as a place to test the theories developed for their reactors. And, for solar observers, lab experiments are filling in details lost across the 150 million km of space between sun and Earth.

"The laboratory work brings an air of reality," solar observer Canfield says. "It really holds your feet to the fire."

Yet another method used by scientists who study the Sun involves the collection of neutrinos, which are subatomic particles without an electric charge. The most sensitive detector of neutrinos is located at the Sudbury Neutrino Observatory (SNO), a facility that is discussed in "Physics Bedrock Cracks, Sun Shines In."

Neutrinos, first discovered in 1956, come in different types, or "flavors." The flavors each have different weights, with muon varieties being heavier than tau neutrinos. While neutrinos come from various sources, one of the largest producers is the Sun, where the subatomic particles are created in reactions involved in the Sun's production of light and energy.

Neutrinos may be players in a cosmic mystery. Most scientists believe that the known mass of objects within the universe, such as that belonging to planets such as Earth and the stars, cannot account for certain processes such as the gravitational pull between galaxies. Some other form of mass must exist. Could neutrinos be a contributing factor? The following article explores that question. —JV

"Physics Bedrock Cracks, Sun Shines In"
by Peter Weiss
Science News, June 23, 2001

By solving a decades-long mystery about the sun, researchers have set off a scientific ripple that could alter conceptions of the universe as a whole.

In the first scientific data to emerge from the Sudbury (Ontario) Neutrino Observatory (SNO) in Canada, physicists have found evidence that their most fundamental theory of the universe—the so-called standard model—contains a major flaw. They've also

uncovered a new clue to the composition of the dark matter thought to make up 95 percent of the mass of the cosmos.

The revelations arise from a new study of the neutrino, an elusive form of matter. Neutrinos come in three types—which scientists call flavors—known as the electron neutrino, muon neutrino, and tau neutrino. These neutrinos span a large range of energies.

Since the 1960s, scientists have measured the sun's copious neutrino output as an indicator of its internal, nuclear-fusion reactions. However, solar-neutrino observatories examining different energy ranges have consistently detected only about half or fewer of the neutrinos that theorists predict they should find. Finally, scientists can now say they know why.

"Our measurements provide an answer to what has been a major puzzle in science for over 30 years," says SNO Director Arthur B. McDonald of Queen's University in Kingston, Ontario.

Solar-fusion reactions unleash only enough energy to make electron neutrinos, so that's the flavor that scientists were expecting to detect. But the new data indicate that electron neutrinos oscillate with other flavors as they make their way to Earth. Because detectors have been less sensitive to muon and tau than to electron neutrinos, the result was a systematic undercounting of solar neutrinos.

"Neutrinos are very confused, very schizophrenic," explains John N. Bahcall of the Institute for Advanced Studies in Princeton, N.J. By pinpointing the source of the neutrino shortfall, the Sudbury team has made "a beautiful measurement" that "confirms our calculations of how the sun shines," he says.

Sudbury scientists unveiled their data on Monday in Victoria, British Columbia, at the annual congress of the Canadian Association of Physicists.

Evidence of neutrinos' schizophrenia first surfaced in 1998. In Japan, physicists were studying muon neutrinos produced by cosmic rays hitting Earth's atmosphere. The researchers at the Super-Kamiokande detector in Kamioka observed that more muon neutrinos arrived at their instrument from directly above the underground device than along other paths through the Earth. The scientists concluded that muon neutrinos taking the longer routes had time to oscillate into other flavors that were more difficult for their detector to see.

These findings shook up particle physics because the standard model rules out such oscillations. What's more, according to quantum mechanics, neutrinos could oscillate only if they had mass—another contradiction to the standard model.

Although the standard model has been a remarkably successful theory, the crack in it that appeared at Super-Kamiokande has now been widened by the SNO finding, researchers say.

Besides studying atmospheric neutrinos, researchers at Super-Kamiokande have measured solar neutrinos. They find 45 percent of the predicted particle flow. The new Sudbury data follow up directly on that measurement by tallying neutrinos in the same energy range. But unique features of the Canadian instrument enable

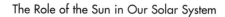

it to reveal details about those neutrinos unattainable with Super-Kamiokande.

The Japanese detector, filled with ordinary water, can't distinguish among the solar-neutrino flavors. However, using 1,000 tons of heavy water as a detector, the Sudbury researchers can measure how many electron neutrinos reach their device, which sits in a working nickel mine.

This measurement has enabled scientists to infer how many electron neutrinos reach Super-Kamiokande. Discerning electron neutrinos from other neutrinos is the key capability that had been needed to solve the solar puzzle, scientists say.

The Sudbury device reveals that the electron neutrinos arriving at an earthly detector total only 35 percent of the predicted solar-neutrino output. Therefore, the additional neutrino flow detected at Super-Kamiokande must be made of up muon and tau neutrinos. Moreover, they must have left the sun as electron neutrinos—the only kind it produces—and changed flavor along the way.

By knowing the sensitivity of Super-Kamiokande to muon and tau neutrinos, the Sudbury team has been able to extrapolate that the total solar output is almost precisely what theorists have called for.

Although the SNO team is still collecting data, the evidence for oscillating solar neutrinos "is pretty definitive at this point," comments Stephen J. Parke of Fermi National Accelerator Laboratory in Batavia, Ill.

Moreover, these new data refine scientists' view of dark matter, says SNO's Kevin T. Lesko of Lawrence Berkeley (Calif.) National Laboratory.

For decades, scientists have known that the mass of the visible objects in the cosmos is far too little to account for features of the universe such as the gravitational tug that keeps galaxy clusters together. Researchers have hypothesized that some combination of familiar, nonluminous matter and some unknown matter—together dubbed dark matter—makes up the difference.

Researchers had suspected that the total neutrino mass was probably not the overwhelming contributor to dark matter, Lesko notes. Now, the Sudbury findings permit researchers for the first time to set a firm limit: The universe's neutrinos make up no more than about half the dark matter.

"Predicting the Sun's Oxygen Isotope Composition" highlights two other methods that scientists use to study the Sun. One method involves the collection of oxygen isotopes, which are oxygen atoms that can have different physical properties, such as different mass. Although oxygen isotopes can be collected from space, many of them appear to be derived from comets and objects other than the Sun.

In September 2004, the Genesis spacecraft, which had collected oxygen isotopes directly from solar wind, experienced a crash landing.

It remains unclear if the isotopes were lost or otherwise contaminated during the landing. Even without such direct samples, scientists use a second method of analysis, called modeling, to predict the Sun's chemistry. Models involve a hypothesis, which is tested against known science that can be verified here on Earth.

The research described in the following article is exciting because it involves the solar nebula. This was the cloud of dust and gas that many scientists believe led to the creation of Earth and the rest of our solar system. The more researchers can learn about elements that exist both on the Sun and on Earth, such as oxygen, and the isotopes associated with these elements, the more we will learn about the origin of the solar system. —JV

"Predicting the Sun's Oxygen Isotope Composition"
by Qing-zhu Yin
Science, September 17, 2004

Most inner solar system materials have distinct oxygen isotopic compositions. A more pronounced variation is observed for some high-temperature components of the most primitive meteorites (see the figure [in the original article]), which represent the most pristine samples of the solar nebula from which our solar system formed. Clearly, the solar system has

^{16}O-rich and ^{16}O-poor reservoirs (1), but exactly what this observation tells us about the early solar nebula remains elusive. Part of the difficulty is that the oxygen isotope composition of the Sun—which represents 99.87% by mass of the solar system—is not known with sufficient precision to determine how the planetary materials evolved from the bulk reservoir.

Recently, a new class of models (2–4) has been proposed to resolve this long-standing problem. In contrast to earlier models that concentrated on the symmetry of minor gas species such as O_3, O_2, and CO_2 (5), the new models focus on photochemical isotopic effects on major volatile species such as CO (second in abundance to H_2 in molecular clouds). On the basis of such a model, Yurimoto and Kuramoto postulate on page 1763 of this issue (4) that oxygen isotope fractionation occurred in the molecular cloud that collapsed to form the solar nebula. They also explain how these isotopic effects may have been transported into the primitive meteorites (called chondrites) of the inner solar system.

The new model (4) tracks a parcel of dust and gas in a molecular cloud with an initially homogeneous oxygen isotopic composition. When the molecular cloud is subjected to ultraviolet (UV) radiation from an external source such as a nearby star, photodissociation of CO occurs. The most abundant isotopomer, ^{12}C^{16}O, quickly consumes all UV photons that have the appropriate energies for dissociating it, and its dissociation therefore stops at the cloud surface. Dissociation of less abundant isotopomers such as ^{12}C^{17}O and ^{12}C^{18}O

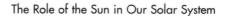

requires UV photons with slightly different energies. Because these photons are not consumed as quickly by the less abundant species, dissociation of the latter continues deeper inside the molecular cloud. This "UV self-shielding" is widely known to occur on the edge of molecular clouds. It generates [17,18]O-rich atomic oxygen and ^{16}O-rich CO inside the molecular cloud.

In the dense, cold core of the molecular cloud, the [17,18]O-rich atomic oxygen rapidly reacts with hydrogen to form a water-ice mantle on silicate grains. The grains retain the original average oxygen isotope composition of the molecular cloud, whereas the mantles are [17,18]O-enriched. When the molecular cloud collapses and the young stellar object ignites, its surroundings are heated through radiation and/or shocks that will chemically and thermally process the ice-mantled grains, producing the observed isotopic heterogeneity along the $\delta^{17}O/\delta^{18}O$ line with slope 1.

Clayton (2) recently suggested that the UV source for self-shielding may have been the nascent Sun itself. Because the nebula was very dense and opaque, UV irradiation is thought to have occurred only very near to the young Sun (7). In this scenario, it is unclear how the chondrite matrix (which, in contrast to some inclusions, was never processed above 400 K) inherited its [17,18]O-rich signature or how most of the water in the solar system became [17,18]O-rich near the Sun and were transported to the outer solar system.

Lyons and Young (3) suggest that CO was dissociated at the disk surface of the solar nebula, further away from the Sun, where the UV source could be either the Sun or other nearby stars. However, it seems that transport of $^{17,18}O$-rich water from the disk surface down to the nebular midplane would have taken too long for the oxygen isotopic signature to be locked into the planetary system. Furthermore, their calculated $\delta^{17}O/\delta^{18}O$ slope of 1.05 to 1.10 (3) is higher than the observed value of 0.94 to 1.00.

The model of Yurimoto and Kuramoto (4) is not subject to such a time constraint, because the $^{17,18}O$-rich water is generated in the molecular cloud well before the solar-nebula disk forms. The close association of the icy mantle with the dust grains readily explains the $^{17,18}O$-rich signatures in the chondrite matrices. However, the authors simply assume that the $\delta^{17}O/\delta^{18}O$ slope is 1 (4). Clearly, the exact production ratio of $^{17}O/^{18}O$ in CO self-shielding as a function of UV fluence needs to be determined in the laboratory to better than 5%, because the astronomical observations are too imprecise.

According to the self-shielding models, the oxygen isotopic composition of the bulk solar nebula (hence the Sun) has $\delta^{17,18}O = -50‰$, similar to some minor constituents of chondrites, such as calcium-aluminum-rich inclusions (CAIs), amoeboid olivine aggregates, and some rare chondrules (most chondrules are ^{16}O-depleted). All other inner solar system materials must then have changed by more than 50‰ from the original composition to reach the current values just above 0‰.

However, some rare chondrules (8) have $^{17,18}O$ as low as -70 to -80, implying that even CAIs and amoeboid olivine aggregates near -50 are not the most pristine

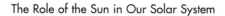

representatives of the bulk solar nebula. This raises the question of whether any existing traces of material retain the oxygen isotope composition of the primordial solar nebula.

The self-shielding model depends strongly on the available UV flux: Too much UV will dissociate all CO, whereas too little will not dissociate it at all. In both cases, no isotopic fractionation occurs. Thus, if the model is correct, the observed oxygen isotopic variations put constraints on the total UV fluence in our solar nebula, which may in turn provide information about the astrophysical settings in which our solar system originated. Isotopic effects in elements other than oxygen, such as hydrogen, nitrogen (9), and sulfur, may also elucidate the astrophysical site where self-shielding occurs.

Clayton originally proposed that the slope 1 line represents mixing of pure ^{16}O (produced in stellar nucleosynthesis by He burning) with the "normal" oxygen isotope mixture typical for most inner solar system materials (1). He now advocates the self-shielding model (2), but the nucleosynthetic origin may still be viable. Observed ^{60}Fe concentrations in chondrites (10, 11) indicate that a supernova must have contributed freshly synthesized material to the solar nebula shortly before the birth of the solar system. One presolar grain with almost-pure ^{16}O ($\delta^{17}O = -921 \pm 47$‰ and $\delta^{18}O = -830 \pm 65$‰) has been found in a chondrite (12); it probably originated in the supernova. The slope 1 line would then constrain the dilution factor of the interstellar medium by the freshly synthesized supernova material to between 0.01 and 5 %.

A verdict on the predicted solar oxygen isotopic composition would have been expected soon had NASA's *Genesis* spacecraft returned the solar wind samples safely to Earth as planned on 8 September 2004. The mission aimed to measure the oxygen isotope composition of the solar wind to better than 1 ‰. However, oxygen isotope measurements on metal grains recovered from lunar soil—exposed to the solar wind for 4 billion years, instead of just 884 days—indicate an extreme ^{17,18}O enrichment of up to 80 ‰ (13), opposite to the prediction of the self-shielding model (2–4). This observation highlights potential challenges to determining the solar oxygen isotopic composition from solar wind samples.

NASA's *Stardust* spacecraft collected interstellar dust particles in May 2000 and 2002. In January 2004, it encountered comet Wild 2 and collected at least 1000 cometary particles. Once returned to Earth in January 2006, these samples may become the most important asset to study the link between interstellar and nebular chemistry. NASA has also just launched the "Oxygen in the Solar System" initiative aimed at deriving an overarching model to elucidate the origins of isotopic complexity. We may not have the final answer to what is arguably the most fundamental outstanding problem in cosmochemistry, but we will soon learn a great deal.

References and Notes

1. R. N. Clayton, *Annu. Rev. Earth Planet. Sci.* 21, 115 (1993).
2. R. N. Clayton, *Nature* 415, 860 (2002) .
3. J. R. Lyons, E. D. Young, *Lunar Planet.* Sci. 35, 1970 (2004).
4. H. Yurimoto, K. Kuramoto, *Science* 305, 1763 (2004).
5. M. H. Thiemens, *Science* 283, 341 (1999).
6. $\delta^{17,18}O = [(^{17,18}O/^{16}O)_{Sample}/(^{17,18}O/^{16}O)_{SMOW} - 1] \times 1000$; (SMOW stands for Standard Mean Ocean Water. The less conventional MC notation used in 4) is converted to $\delta^{17,18}O_{SMOW}$ notation with $\delta^{17,18}O_{SMOW} = \delta^{17,18}O_{MC} - 50\,‰$.
7. F. H. Shu, H. Shang, T. Lee, *Science* 271, 1545 (1996).
8. S. Kobayashi, H. Imai, H. Yurimoto, *Geochem. J.* 37, 663 (2003).
9. R. N. Clayton, Meteorit. *Planet. Sci.* 37 (Suppl.), 35 (2002).
10. S. Tachibana, G. R. Huss, *Astrophys. J.* 588, L41 (2003).
11. S. Mostefaoui, G. W. Lugmair, P. Hoppe, A. El Goresy, *New Astron.* Rev. 48, 155 (2004).
12. L. R. Nittler, C. M. O. Alexander, J. Wang, X. Gao, *Nature* 393, 222 (1998).
13. T. R. Ireland, P. Holden, M. Norman, J. Clarke, *Lunar Planet. Sci.* 35, 1448 (2004).
14. E. D. Young, S. S. Russell, *Science* 282, 452 (1998).

Reprinted with permission from Yin, Qing-zhu, "Predicting the Sun's Oxygen Isotope Composition," SCIENCE 305:1729-1730 (2004). © 2004 AAAS.

Solar Wind, Magnetism, and Polarity

2

In terms of Earth's weather, wind is the natural movement of air in a horizontal direction, but for the Sun, "wind" has a somewhat different meaning. Usually, it is referred to as solar wind, which consists of protons and electrons that flow outward from the Sun in all directions. The only real similarity between solar wind and wind on Earth is that both have motion and create an energized force. If you have ever experienced a hurricane on Earth, you know how forceful wind can be.

Solar wind is so powerful that the particles within it come very close to reaching Earth. In fact, they would flow over us all of the time, but Earth has a magnetic field that helps to prevent the particles from reaching the surface of our planet. Earth's field is called the magnetosphere. It is a region with electrically charged particles that behave according to the rules of magnetic electricity rather than to the fundamental rules of gravity that define structures on Earth.

The paper "Transport of Solar Wind into Earth's Magnetosphere Through Rolled-Up Kelvin–Helmholtz Vortices" discusses research on what happens when solar wind enters Earth's magnetosphere. The scientists found that certain conditions lead to instabilities. These instabilities, in turn, produce variations in space weather. Like the difference in terminology between "wind" and "solar wind," "space weather" differs from weather here on Earth in that it refers to magnetic changes, as opposed to rain, snow, air wind, and so on. Like a television news weatherman, researchers such as H. Hasegawa and colleagues attempt to predict, and to better understand, space weather. —JV

"Transport of Solar Wind into Earth's Magnetosphere Through Rolled-Up Kelvin–Helmholtz Vortices"
by H. Hasegawa, M. Fujimoto, T. D. Phan, H. Rème, A. Balogh, M. W. Dunlop, C. Hashimoto, and R. TanDokoro
Nature, August 2004

Along the outer boundary of Earth's magnetosphere, there is a boundary layer that contains plasma of dominantly solar-wind origin[6] (Fig. 1a [in the original article]). The boundary layer exists for all orientations of the solar-wind magnetic field, but it tends to be thicker when the solar-wind field points northward.[3, 4] The existence of the boundary layer implies penetration of solar wind across the magnetopause. Although reconnection between solar-wind and terrestrial magnetic fields can readily account for solar-wind entry during southward solar-wind magnetic field conditions, it is at present not known how the plasma crosses the magnetopause when the solar-wind magnetic field is oriented northward and parallel to geomagnetic fields and reconnection is less efficient, although there is a suggestion that simultaneous northern and southern cusp reconnection could result in the formation of the boundary layer at low latitudes.[7] Several candidate local entry mechanisms unrelated to reconnection have been proposed,[8] one of which is the Kelvin–Helmholtz instability (KHI) that could occur along the flanks of the magnetosphere where the shocked solar wind is flowing fast relative to the stagnant magnetospheric plasma[9, 10] (Fig. 1 [in the original article]). Recent numerical simulation models[11–16] suggest that fast plasma transport across the magnetopause can be accomplished by the KHI only when the KHI has grown sufficiently to form rolled-up vortices that can engulf plasmas from both sides of the magnetopause. In these models, the collapse of, or reconnection within, such a vortex (in the nonlinear phase of the KHI) is responsible for the plasma transport.

Multiple and quasi-periodic encounters by spacecraft with the magnetopause and vortex-like flow perturbations near the magnetopause have been reported, and are often interpreted as representing surface waves or vortices excited by the KHI.[17–21] But as long as these signatures are observed only by a single spacecraft, one cannot tell unambiguously whether the KHI

has reached its nonlinear stage and is generating rolled up vortices, which are the crucial ingredient for plasma transport, or if they are just ripples or small-amplitude Kelvin–Helmholtz (KH) vortices on the magnetopause surface. The KH rolled-up vortex expected to form at the magnetopause has complex structures, such as vortical plasma flow and a filament-like high density region intruding into the low density (magnetospheric) region (See Fig. 1 [in the original article]). To resolve such complex structures in the KH vortices, multipoint *in situ* measurements as carried out by the Cluster mission are essential, as is comparison with realistic three-dimensional (3D) plasma simulations.

Here we report multi-spacecraft measurements that provide unambiguous evidence for rolled-up vortices at the flank magnetopause as well as simultaneously observed boundary-layer characteristics that result from plasma transport across the boundary, such as would be expected from the suggested KHI mechanism.

The four Cluster spacecraft forming a tetrahedron made a fortuitous direct encounter with the rolled-up vortices (Fig. 2 [in the original article]) on 20 November 2001 when the upstream solar-wind magnetic field observed by the ACE spacecraft pointed northward, that is, when reconnection was less efficient but the condition was favourable for the KHI[22] at the low-latitude magnetopause. The solar-wind and magnetospheric magnetic fields on the dusk flank magnetopause were approximately parallel throughout the 16 min interval (Fig. 2f, g [in the original article]). The period of the vortices encounter was embedded in a more than 13-h interval of quasi-periodic plasma and magnetic field perturbations related to deformations of the magnetopause (09:30–23:30 UT). Rolled-up vortices were identified far more clearly by taking full advantage of multi-spacecraft information (Fig. 2e [in the original article]). The flow vectors, transformed into the frame of the vortices and viewed from the north, rotate anticlockwise around the centre of the vortices (marked by the red vertical lines in Fig. 2 [in the original article]), as expected at the duskside magnetopause. Near these vortices, the density observed by the Cluster 1 spacecraft (C1) located farthest from the magnetopause (marked by the blue arrows in Fig. 2b [in the original article]) toward the magnetosphere was often higher than that observed by C3 or C4 (Fig. 2c, d [in the original article]). These instances of higher density at C1 are marked by red bars in Fig. 2c [in the original article]. Such high density regions appear to be connected to the dense solar wind on the anti-sunward side, for example, at 20:27:30 UT in Fig. 2d [in the original article] rather than detached from the solar-wind region. This feature is consistent with the simulation result (Fig. 1b [in the original article]), and indicates that the dense regions result from the roll-up of the solar-wind plasma associated with the growth of the KHI (Fig. 1b [in the original article]), not from the impulsive penetration process.[23] The density variation observed by C1 (Fig. 2c [in the original article]) is similar to that expected when a synthetic spacecraft moves through the central portion of the simulated vortex and crosses the magnetopause back and forth (Fig. 1b [in

the original article]), suggesting that C1 was in the vicinity of the centre of the vortices.

In addition to the density and flow signatures of the rolled-up KH vortices, we identified a unique magnetic field perturbation pattern associated with these vortices, which should appear only in a 3D configuration of the magnetosphere where the KH-unstable plasma sheet is sandwiched between the KH-stable northern and southern lobes (Fig. 1a [in the original article]). A numerical simulation of the KHI that considers this 3D magnetosphere-like geometry predicts that this field perturbation manifests in boundary regions between the plasma sheet and the lobes. This is because only low-latitude portions of the field lines surrounding the magnetopause are engulfed into the vortices, while those at high latitudes are unaffected (See Fig. 1 legend [in the original article]). Fig. 2f [in the original article] shows that magnetic field perturbances seen in the high density (solar-wind) region ($\triangle\beta_x > 0$ and $\triangle\beta_y < 0$) have polarities in precise agreement with the 3D KHI effect on the magnetic field on the south side of the equatorial plane where Cluster resided (Fig. 1a [in the original article]). The combined plasma and magnetic field observations provide unambiguous identification of rolled-up vortices at the magnetopause, which, according to theory, is a key (and necessary) ingredient for plasma transport via KHI.

Evidence for plasma transport across the magnetopause was indeed observed. Cold solar-wind and hot magnetospheric ion populations are found to coexist in the vortices (Figs. 2a, 3 [in the original article]). Significant amounts of the solar-wind (<2 keV) and

magnetospheric (>5 keV) ions were detected simultaneously in the same region on the magnetosphere side of the magnetopause. The appearance of rolled-up vortices in the vicinity of the boundary layer is strongly suggestive of the KHI mechanism for plasma transport across the boundary.

We could not deduce the exact microphysical process that leads to plasma transport within the KH vortices on the basis of the present measurements. However, we could rule out local reconnection, because during the interval of the vortex observations, we did not find signatures of plasma acceleration due to magnetic stresses[24, 25] and D-shaped ion distributions characteristic of reconnection[26] (See Supplementary Fig. 1 [in the original article]). To exclude, conclusively, the possibility of remote (high-latitude) reconnection supplying the plasma observed at low latitude is more difficult. We do note, however, that the boundary layer ions, detected off the equator, were flowing poleward in precisely the same direction as the solar-wind flow. This seems inconsistent with the idea that high-latitude magnetopause reconnection,[27] which would result in an equatorward flow at the observation point, produced the observed boundary layer. These observations thus indicate that reconnection occurred neither locally nor at the high-latitude magnetopause during the time of the observations of KH vortices. However, we cannot rule out the possibility that reconnection occurred in the past (but had ceased to exist) to produce a boundary layer which is now rolled-up by the KHI. Proving the existence of such a scenario is, however, even more difficult.

We estimate the speed of motion of the vortices by averaging over the vortices' interval the velocity values measured by the three spacecraft (C1, C3 and C4) on which the plasma instruments were operative. Time is then translated into distance of the spacecraft from a certain position in a vortex, and the length scale of one vortex is estimated to be 40,000–55,000 km (See Fig. 2 [in the original article]). Consequently, the initial thickness of the velocity shear layer is inferred to have been roughly 5,000–7,000 km, because the wavelength of the fastest growing KH mode is approximately eight times the initial total thickness of the velocity shear layer.[28] According to numerical simulations, the width of the sufficiently developed KH vortex equivalent to that of the plasma boundary layer reaches about four times the initial thickness. We therefore infer that the boundary layer with the thickness of 20,000–28,000 km had been formed in the most KH-unstable low-latitude regions near, or at least further downstream of, the observation site.

The present results indicate that the KHI occurs at the flank magnetopause and that it may lead to solar-wind entry, perhaps via non-reconnection-associated processes, for northward solar-wind magnetic field. But the microphysical process that causes the plasma transport in the KH vortices, which would control the rate at which mass and energy of the solar wind are transferred, remain to be understood. The identification of the transport processes requires high-resolution measurements capable of resolving small-scale structures and dynamics embedded in the vortices. Such observations (which will be carried out by the future NASA Magnetospheric Multiscale mission[29] and the Japanese SCOPE mission) would enable us to determine the relative contributions of reconnection and the KHI during northward solar-wind magnetic field conditions.

1. Akasofu, S.-I. Energy coupling between the solar wind and the magnetosphere. *Space Sci. Rev.* 28, 121–190 (1981).

2. Cowley, S. W. H. in *Magnetic Reconnection in Space and Laboratory Plasmas* (ed. Hones, E. W.) 375–378 (Geophys. Monograph 30, American Geophysical Union, Washington DC, 1984).

3. Mitchell, D. G. *et al.* An extended study of the low-latitude boundary layer on the dawn and dusk flanks of the magnetosphere. *J. Geophys. Res.* 92, 7394–7404 (1987).

4. Hasegawa, H., Fujimoto, M., Saito, Y. & Mukai, T. Dense and stagnant ions in the low-latitude boundary region under northward interplanetary magnetic field. *Geophys. Res. Lett.* 31, L06802 (2004).

5. Chandrasekhar, S. *Hydrodynamic and Hydromagnetic Stability* (Oxford Univ. Press, New York, 1961).

6. Sckopke, N. G. *et al.* Structure of the low-latitude boundary layer. *J. Geophys. Res.* 86, 2099–2110 (1981).

7. Song, P. & Russell, C. T. Model of the formation of the low-latitude boundary-layer for strongly northward interplanetary magnetic-field. *J. Geophys. Res.* 97, 1411–1420 (1992).

8. Sibeck, D. G. *et al.* in *Magnetospheric Plasma Sources and Losses* Ch. 5 (ed. Hultqvist, B.) 207–283 (Space Sciences Series of ISSI 6, Kluwer Academic, Dordrecht, 1999).

9. Dungey, J. W. in *Proc. Ionosphere Conf.* 225 (Physical Society of London, 1955).

10. Miura, A. Anomalous transport by magnetohydrodynamic Kelvin-Helmholtz instabilities in the solar wind-magnetosphere interaction. *J. Geophys. Res.* 89, 801–818 (1984).

11. Thomas, V. A. & Winske, D. Kinetic simulations of the Kelvin-Helmholtz instability at the magnetopause. *J. Geophys. Res.* 98, 11425–11438 (1993).

12. Fujimoto, M. & Terasawa, T. Anomalous ion mixing within an MHD scale Kelvin-Helmholtz vortex. *J. Geophys. Res.* 99, 8601–8613 (1994).

13. Huba, J. D. The Kelvin-Helmholtz instability: Finite Larmor radius magneto-hydrodynamics. *Geophys. Res. Lett.* 23, 2907–2910 (1996).

14. Nykyri, K. & Otto, A. Plasma transport at the magnetospheric boundary due to reconnection in Kelvin-Helmholtz vortices. *Geophys. Res. Lett.* 28, 3565–3568 (2001).

15. Matsumoto, Y. & Hoshino, M. Onset of turbulence induced by a Kelvin-Helmholtz vortex. *Geophys. Res. Lett.* 31, L02807 (2004).
16. Nakamura, T. K. M., Hayashi, D., Fujimoto, M. & Shinohara, I. Decay of MHD-scale Kelvin-Helmholtz vortices mediated by parasitic electron dynamics. *Phys. Rev. Lett.* 92, 145001 (2004).
17. Ogilvie, K. W. & Fitzenreiter, R. J. The Kelvin-Helmholtz instability at the magnetopause and inner boundary layer surface. *J. Geophys. Res.* 94, 15113–15123 (1989).
18. Kivelson, G. K. & Chen, S.-H. in *Physics of the Magnetopause* (eds Song, P., Sonnerup, B. U. Ö. & Thomsen, M. F.) 257–268 (Geophys. Monograph 90, American Geophysical Union, Washington DC, 1995).
19. Hones, E. W. Jr *et al.* Further determination of the characteristics of magnetospheric plasma vortices. *J. Geophys. Res.* 86, 814–820 (1981).
20. Fairfield, D. H. *et al.* Geotail observations of the Kelvin-Helmholtz instability at the equatorial magnetotail boundary for parallel northward fields. *J. Geophys. Res.* 105, 21159–21173 (2000).
21. Fujimoto, M., Tonooka, T. & Mukai, T. in *Earth's Low-latitude Boundary Layer* (eds Newell, P. T. & Onsager, T.) 241–251 (Geophys. Monograph 133, American Geophysical Union, Washington DC, 2003).
22. Miura, A. Dependence of the magnetopause Kelvin-Helmholtz instability on the orientation of the magnetosheath magnetic field. *Geophys. Res. Lett.* 22, 2993–2996 (1995).
23. Lemaire, J. Impulsive penetration of filamentary plasma elements into magnetospheres of Earth and Jupiter. *Planet. Space Sci.* 25, 887–890 (1977).
24. Paschmann, G. *et al.* Plasma acceleration at the Earth's magnetopause: Evidence for reconnection. *Nature* 282, 243–246 (1979).
25. Phan, T.-D. *et al.* Extended magnetic reconnection at the Earth's magnetopause from detection of bi-directional jets. *Nature* 404, 848–850 (2000).
26. Fuselier, S. A. in *Physics of the Magnetopause* (eds Song, P., Sonnerup, B. U. Ö. & Thomsen, M. F.) 181–187 (Geophys. Monograph 90, American Geophysical Union, Washington DC, 1995).
27. Kessel, R. L. *et al.* Evidence of high-latitude reconnection during northward IMF: Hawkeye observations. *Geophys. Res. Lett.* 23, 583–586 (1996).
28. Miura, A. & Pritchett, P. L. Nonlocal stability analysis of the MHD Kelvin-Helmholtz instability in a compressible plasma. *J. Geophys. Res.* 87, 7431–7444 (1982).
29. *Report of the NASA Science and Technology Definition Team for the Magnetospheric Multiscale (MMS) Mission* (NASA/TM-2000-209883, Goddard Space Flight Center, Greenbelt, MD, 1999).

Another term that Sun scientists use is "magnetopause." It refers to the outer boundary of Earth's magnetosphere, or any magnetosphere. Just as the word suggests, the boundary is where a pause occurs when solar wind encounters it. Sometimes solar wind passes through the boundary, and other times it does not. Research shows that the connection between solar wind and Earth's magnetosphere is intermittent.

The paper "Continuous Magnetic Reconnection at Earth's Magnetopause" explores the forces driving the connection. Is our magnetosphere controlling the activity, or is it controlled by the ever-changing solar wind? Through extensive modeling, the researchers discovered that the interface between solar wind and Earth's magnetopause appears primarily to be fueled by the solar wind. One piece of evidence consists of the dayside proton auroral spot in the ionosphere. Auroras are spectacular emissions of light emitted by atoms that have been excited by electrons. The ionosphere is a place in Earth's atmosphere where ionization, or conversion of atoms into ions, of atmospheric gases takes place. The proton auroral spot leaves a visual signature that, in part, enabled H. U. Frey and colleagues to theorize about what happens at the magnetopause. —JV

"Continuous Magnetic Reconnection at Earth's Magnetopause"
by H. U. Frey, T. D. Phan, S. A. Fuselier, and S. B. Mende
Nature, December 4, 2003

The most important process that allows solar-wind plasma to cross the magnetopause and enter Earth's magnetosphere is the merging between solar-wind and terrestrial magnetic fields of opposite sense-magnetic reconnection.[1] It is at present not known whether reconnection can happen in a continuous fashion or whether it is always intermittent. Solar flares[2] and magnetospheric substorms[3]—two phenomena believed to be initiated by reconnection—are highly burst-like occurrences, raising the possibility that the reconnection process is intrinsically intermittent, storing and releasing magnetic energy in an explosive and uncontrolled manner. Here we show that reconnection at Earth's high-latitude magnetopause is driven directly by the solar wind, and can be continuous and even quasi-steady over an extended period of time. The dayside proton auroral spot in the ionosphere—the remote signature of high-latitude magnetopause reconnection[4]—is present continuously for many hours. We infer that reconnection is not intrinsically intermittent; its steadiness depends on the way that the process is driven.

Models of reconnection are typically steady state. They describe what happens once the process is initiated and allowed to proceed indefinitely. Transient reconnection models have also been developed, which describe the consequence of artificially varying the reconnection rate, even including times when the rate drops to zero.[5] A key unanswered question, both theoretically and observationally, is how long the process can maintain itself naturally once initiated.[6] In other words, is the process intrinsically intermittent or continuous? Intermittent reconnection turns on and off. Continuous reconnection operates at a variable rate but never ceases; if the fluctuation is a small fraction of the average then the reconnection is classed as "quasi-steady." Attempts to deduce the steadiness of reconnection on the basis of *in situ* magnetopause observations were inconclusive.[7] Because signatures of reconnection (for example, plasma jets) are localized in the thin magnetopause, they can be observed only for a short time (a few minutes) when a spacecraft intersects the layer, even if the reconnection is actually continuous. Much of the published literature from ground-based or *in situ* measurements in the cusp has reported intermittent reconnection.[8-12]

It has been established recently that the dayside proton auroral spot[13, 14] is created directly by solar-wind protons leaking through, and accelerated at, the magnetopause by way of reconnection between solar-wind and magnetospheric magnetic fields[4] (Fig. 1 [in the original article]). The proton aurora spot is detected by the SI-12 imager on the IMAGE spacecraft.[15] It detects the Doppler-shifted Lyman-α photons corresponding to precipitating charge-exchanged protons with energies of a few keV. As this is not the average energy of a typical proton distribution in the shocked solar wind, some sort of energization process is required. Magnetic reconnection at the magnetopause readily provides this energization.[4] Thus, the SI-12 imager is well suited to provide global images of the aurora created from precipitating protons that have been energized by magnetic reconnection at an active reconnection site.

Here we report two notable examples of long-duration proton auroral spot observations that imply continuous and even quasi-steady magnetopause reconnection for many hours. These observations were made during northward solar-wind magnetic field conditions, when the reconnection site is expected to be at the high-latitude magnetopause where the solar-wind and magnetospheric magnetic fields are antiparallel (Fig. 1 [in the original article]).

The first reconnection event occurred on 18 March 2002 (Fig. 2 [in the original article]). For the entire ~4-h interval shown, the solar-wind dynamic pressure (Fig. 3a [in the original article]) was constant (~17 nPa). The solar-wind magnetic field (B, Figure 3b [in the original article]) was persistently northward directed ($B_z > 0$) while its y (east–west) component (B_y) fluctuated between positive and negative. Figures 2 and 3c–e [in the original article] show the uninterrupted presence of a bright proton auroral spot during this ~4-h interval. During a brief interval (5 min), the Cluster spacecraft by chance crossed the high latitude magnetopause and observed proton jets accelerated by reconnection at the magnetopause.[4] The reconnection jets were observed on field lines that are linked to the spot in the ionosphere and with energy fluxes consistent with the spot brightness, thus providing direct evidence that the spot represents the remote signature of high-latitude magnetopause reconnection. Although Cluster could only observe reconnection for 5 min, the uninterrupted presence of the spot in the images in Fig. 2 [in the original article] implies continuous reconnection over ~4h. Note that the peak and average brightness of the spot (Fig. 3c [in the original article]) remain high over the whole observation period, with some fluctuations.

After 14:40 UT, the magnetic local time (MLT) location of the spot (Figs. 2, 3c [in original article]) follows very closely the changes in the y component of the solar-wind magnetic field (Fig. 3b [in the original article]), with spot locations at pre-noon MLT for negative B_y and with post-noon locations with positive B_y (ref. 16). The correlation implies that reconnection is a directly driven process at the magnetopause and is consistent with antiparallel reconnection. That is, the reconnection site is located where the internal magnetospheric magnetic field is antiparallel to the external shocked solar-wind field.[14, 17] As the solar-wind magnetic field changes, the antiparallel reconnection site moves on the magnetopause. This is seen in the ionosphere as motion in MLT but not in latitude. This motion, combined with the near-constant brightness of the spot, indicates that the reconnection site is continuously active (not intermittent). Thus, viewed globally, the reconnection process never stops at the high-latitude magnetopause. Finally, as the spot moves in MLT in response to the solar wind B_y, it does not leave a long trail in latitude or in MLT, which would have indicated that proton aurorae are still created long (more than 4–5 min) after the cessation of reconnection.

The second continuous reconnection event occurred on 17/18 September 2000 (Fig. 4 [in the original article]. This event does not have the added benefit of simultaneous *in situ* observations at the magnetopause. However, it is remarkable in that the dayside proton aurora spot was observed uninterruptedly for ~9 h (01:00–10:00 on 18 September), in addition to more than 3 h of continuous presence of the bright proton spot during the previous IMAGE apogee (northern hemisphere) pass (16:20–19:30 UT on 17 September). Before 05:45 UT on 18 September, similar to the previous event, the MLT location of the spot is well correlated with changes in the y-component of the solarwind magnetic field, with similar implications in terms of antiparallel reconnection. Between 05:45 and 09:00 UT on 18 September, however, the solar-wind pressure and magnetic field direction were nearly constant. The proton spot brightness and location during the corresponding interval were also remarkably stable. Furthermore, the poleward boundary of the proton aurora spot (red line in Fig. 4d [in the orginal article]) was also rather stable. This stability rules out bursts of reconnection with periods longer than 4–5 min (the images are obtained on a two-minute cadence). Longer periods of reconnection bursts would produce observable brightness and size changes, slow periodic equatorward motions, and sudden poleward jumps of the proton spot poleward boundary.

The above observations can be summarized and further interpreted as follows. The dayside proton auroral spot, which recently has been shown to be the footprint of high-latitude reconnection, was observed continuously for close to 4 h in one event and more than 9 h in another. The continuous presence of the spot even as the solar-wind condition changed implies that continuous magnetic reconnection was occurring over these time spans. In a global sense, it is significant that reconnection at the high-latitude magnetopause never stops for any period of time longer than 4-5 min. For more stable solar-wind conditions in the second half of the second event, reconnection is continuous not only in a global sense, but its rate appears quasi-steady at the local magnetopause as well. The present finding explains the persistence of optical signatures of the cusp from ground-based observations during stable solar-wind magnetic field conditions.[18–20]

For low-latitude reconnection during southward solar-wind magnetic field conditions, it was reported that the proton aurora could be seen on each reconnected field line for up to 5 min following reconnection.[21] In the present high-latitude reconnection case, this time might be longer. Any intermittency in the reconnection rate on 5 min or longer timescales should create observable signatures (changes in spot brightness and latitude) in the proton aurora image sequences. Such changes are not seen in our 2 min resolution images. Thus our findings are in contrast to reports favouring intermittent reconnection, which were made mostly under southward solar-wind magnetic field conditions.[8–10, 12] This contrast is more surprising considering that reconnection is thought to be more intermittent during northward solar-wind

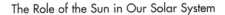

magnetic field, when the reconnection site is located at the highlatitude magnetopause (as opposed to a low-latitude site for southward solar-wind magnetic field) owing to the presence of fast shocked solar-wind flows.[22] A possible explanation for this discrepancy is the fact that most of the previous studies of reconnection have been made locally using ground-based or *in situ* measurements, which could not distinguish temporal variations of reconnection from motion of the reconnection site if the solar-wind magnetic field is not steady. At the local magnetopause, reconnection could well be intermittent if the solar-wind field changes.

The present results suggest that reconnection for northward solar-wind magnetic field is a directly driven process that is continuous when viewed in a global sense, or even in a local sense when the solar-wind conditions are steady. This regime of reconnection is in stark contrast to the storage and release of magnetic energy in explosive reconnection at the Sun (associated with solar flares) or in the magnetotail (associated with magnetospheric substorms). This contrast may suggest that, if driven persistently—in this case the shocked solar wind impinges on the magnetopause continuously—the process can be continuous. To evaluate fully the controlling factors that dictate the temporal behaviour of reconnection is a major objective of the planned NASA Magnetospheric Multiscale mission.[6]

1. Cowley, S. W. H. in *Magnetic Reconnection in Space and Laboratory Plasmas* (ed. Hones, E. W.) 375–378 (Geophysics Monograph 30, American Geophysical Union, Washington DC, 1984).

2. Hudson, H. & Ryan, J. High-energy particles in solar flares. *Annu. Rev. Astron. Astrophys.* 33, 239–282 (1995).

3. Angelopoulos, V. *et al.* Statistical characteristics of bursty bulk flow events. *J. Geophys. Res.* 103, 21257–21280 (1994)

4. Phan, T. *et al.* Simultaneous Cluster and IMAGE observations of cusp reconnection and auroral proton spot for northward IMF. *Geophys. Res. Lett.* 30, doi:10.1029/2003GL016885 (2003).

5. Scholer, M. Magnetic flux transfer at the magnetopause based on single X line bursty reconnection. *Geophys. Res. Lett.* 15, 291–294 (1988).

6. *Report of the NASA Science and Technology Definition Team for the Magnetospheric Multiscale (MMS) Mission* (NASA/TM-2000-209883, Goddard Space Flight Center, Greenbelt, MD, 1999).

7. Phan, T. D. *et al.* Extended magnetic reconnection at the Earth's magnetopause from detection of bi-directional jets. *Nature* 404, 848–850 (2000).

8. Farrugia, C. J., Sandholt, P. E., Denig, W. F. & Torbert, R. B. Observation of a correspondence between poleward moving auroral forms and stepped cusp ion precipitation. *J. Geophys. Res.* 103, 9309–9315 (1998).

9. Lockwood, M. *et al.* Cusp ion steps, field-aligned currents and poleward moving auroral forms. *J. Geophys. Res.* 106, 29555–29569 (2001).

10. Lockwood, M., Davis, C. J., Onsager, T. G. & Scudder, J. D. Modelling signatures of pulsed magnetopause reconnection in cusp ion dispersion signatures seen at middle altitudes. *Geophys. Res. Lett.* 25, 591–594 (1998).

11. Sandholt, P. E. *et al.* Dynamic cusp aurora and associated pulsed reverse convection during northward interplanetary magnetic field. *J. Geophys. Res.* 105, 12869–12894 (2000).

12. Milan, S. E., Lester, M., Greenwald, R. A. & Sofko, G. The ionospheric signature of transient dayside reconnection and the associated pulsed convection return flow. *Ann. Geophys.* 17, 1166–1171 (1999).

13. Frey, H. U. *et al.* Proton aurora in the cusp. *J. Geophys. Res.* A 107, doi:10.1029/2001JA900161 (2002).

14. Fuselier, S. A. *et al.* Cusp aurora dependence on IMF Bz. *J. Geophys. Res.* A 107, doi:10.1029/ 2001JA900165 (2002).

15. Mende, S. B. *et al.* Far ultraviolet imaging from the IMAGE spacecraft: 3. Spectral imaging of Lyman alpha and OI 135.6 nm. *Space Sci. Rev.* 91, 287–318 (2000).

16. Newell, P. T., Meng, C.-I., Sibeck, D. G. & Lepping, R. Some low-altitude cusp dependencies on the interplanetary magnetic field. *J. Geophys. Res.* 94, 8921–8927 (1989).

17. Frey, H. U., Mende, S. B., Fuselier, S. A., Immel, T. J. & Ostgaard, N. Proton aurora in the cusp during southward IMF. *J. Geophys. Res.* A 108, doi:10.1029/2003JA009861 (2003).

18. Mende, S. B., Rairden, R. L., Lanzerotti, L. J. & Maclennan, C. G. Magnetic impulses and associated optical signatures in the dayside aurora. *Geophys. Res. Lett.* 17, 131–134 (1990).

19. Sandholt, P. E. IMF control of polar cusp and cleft auroras. *Adv. Space Res.* 8, 21–34 (1988).
20. Newell, P. T. Do the dayside cusps blink? *Rev. Geophys. Suppl.* 33, 665–668 (1995).
21. Lockwood, M. *et al.* IMF control of cusp proton emission intensity and dayside convection: Implications for component and anti-parallel reconnection. *Ann. Geophys.* 21, 955–982 (2003).
22. Cowley, S.W. H. & Owen, C. J. A simple illustrative model of open flux tube motion over the dayside magnetopause. *Planet. Space Sci.* 37, 1461–1475 (1989).

Magnetism and polarity are two key properties of science that occur on the Sun. An understanding of how a common bar magnet works can help demystify magnetism and polarity. A bar magnet has two sides, or poles. Each pole possesses an attracting force. If metal shavings were to be placed near the magnet, they would fan out in a C shape on either side of the magnet. A similar effect occurs with planets. On Earth, the North Magnetic Pole and the South Magnetic Pole somewhat represent the two ends of a magnet within or around Earth's axis. The polarity of these areas does not significantly change.

On the Sun, however, polarity does change. In general, that likely is due to the fact that the Sun consists of moving gas plasma. In "Coronal Mass Ejections and Solar Polarity Reversal," the authors explore the relationship between solar polarity reversals, which occur on a cyclical basis, and coronal mass ejections. The word "corona" means "crown," and it refers to the outermost layer of the Sun. Mass ejections from this part of the Sun consist of huge eruptions of material from the corona that jet out into interplanetary space.

Coronal mass ejections, as the paper reveals, seem to be linked to solar polarity reversal. While the exact relationship remains unclear, studies such as this one add to our knowledge of solar polarity, which could lead to a better understanding of Earth's polarity and the magnetic properties of other planets and stars. —JV

From "Coronal Mass Ejections and Solar Polarity Reversal"
by N. Gopalswamy, A. Lara, S. Yashiro, and R. A. Howard
The Astrophysical Journal, November 20, 2003

1. Introduction

Common signatures of magnetic polar reversals on the Sun are the disappearance and reformation of polar coronal holes (Webb, Davis, & McIntosh 1984; Bilenko 2002; Harvey & Recely 2002) and the disappearance of the polar crown filaments (PCFs) following a sustained march to the poles (Waldmeier 1960; Makarov, Tlatov, & Sivaraman 2001). These signatures are, of course, related: during the interval between the disappearance of the old-polarity coronal hole and the reformation of the new-polarity coronal hole, the PCFs that approach the poles need to disappear. Filaments/prominences do not occur naked but have overlying closed field structures, commonly known as helmet streamers. Thus, we would expect the complete disappearance of these closed field structures when the polarity changes. Prominence disappearances are known to be either eruptive or thermic (see, e.g., Wagner 1986). Only eruptive prominences can be relevant, because thermic disappearances represent a temporary heating of the prominence. Gopalswamy et al. (2003a) noted that prominence eruptions at high latitudes subsided around the time of the polarity reversals. Since eruptive prominences are almost always accompanied by coronal mass ejections (CMEs), one can identify the disappearance of PCFs with high-latitude (HL) CMEs (Gopalswamy et al. 2003b). Thus CMEs may be intimately connected to the mechanism of polarity reversal. We test this possibility using the well-observed CMEs of solar cycles 23 and 21.

2. Analysis and Results

The primary CME data needed for this study were obtained by the *Solar and Heliospheric Observatory* (*SOHO*) mission's Large Angle and Spectrometric Coronagraph Experiment (LASCO; Brueckner et al. 1995) for cycle 23. The CME rates for 1996–2002 were derived from the online catalog. There were no space-borne coronagraphs operating during the polarity reversal times of solar cycle 22. For cycle 21, the CMEs were observed by the Solwind coronagraph on board the *P78-1* satellite (Cliver et al. 1994). We use the National Solar Observatory's Kitt Peak magnetograph measurements available online to track the evolution of the photospheric magnetic field strength in the polar regions and identify the epochs of polarity reversal. We extend the list of prominence eruptions from the Nobeyama radioheliograph studied by Gopalswamy et al. (2003b) to include events from 2002.

2.1. Prominence Eruptions and CMEs

In a recent paper, Gopalswamy et al. (2003b) established that prominence eruptions (PEs) and CMEs are closely related and had similar latitude dependence. Figure 1 [in the original article] compares the latitudinal distance of PEs and CMEs along with the tilt angle (maximum excursions of the heliospheric current sheet as available from the Wilcox Solar Observatory Web site). We note the close similarity between the PE and CME distributions at various latitudes. Of particular importance for this Letter is the north-south asymmetry in the HL activities. The epochs when the HL PEs and CMEs subside (marked by the vertical lines) are clearly different in the north and south. Nobeyama radioheliograph observes the Sun only for 8 hr day^{-1} and only limb events are automatically detected, so the sample size is relatively small. We show that the epochs of cessation of HL PEs also mark when the general population of HL CMEs subsides.

2.2. CME Occurrence Rate

Figure 2 [in the original article] shows the occurrence rate of LASCO CMEs averaged over Carrington rotation (CR) periods. The error bars are computed based on *SOHO* down times during each rotation (Gopalswamy et al. 2003). Also shown are the daily sunspot numbers (SSNs). Although there is an overall similarity between the SSN and the CME rates, there are clear differences in detail. The CME rate peaks in 2002, roughly 2 yr after the peak in the SSN. The sunspot activity is confined to the active region belt (low to mid latitudes), but the CME activity occurs at all latitudes. Separating HL and low-latitude (LL) CMEs thus provides a convenient way of grouping sunspot-related and PCF-related CMEs. The CME latitudes can be obtained by converting the observed central position angle to the latitude of the CMEs. It is likely that some LL CMEs may be misidentified as HL CMEs because of projection effects. To minimize this, we have grouped CMEs with apparent latitude $\leqslant 40°$ as LL CMEs and those with latitude $\geqslant 60°$ as HL CMEs. The rates of LL and HL CMEs are also shown in Figure 2 [in the original article]. The overall ratio of HL to LL rates is 25 % (20 % when CMEs above and below 60° latitude are considered). Occasionally, the ratio was close to 100 % during short intervals. The LL CME rate is remarkably flat during solar maximum (except for the fluctuations) compared to all the other rates. The HL CME rate displays more variability and is of interest for this study.

2.3. HL CME Rate and Polarity Reversal

The variation of the photospheric magnetic field of cycle 23, averaged over longitudes and poleward of 70°, is shown in Figure 3 [in the original article]. Starting in the middle of the year 1999, the field strengths at both poles decline and the first signs of reversal occur in early 2000.

The reversal is obviously not a sharp process and is completed only after a few episodes of temporary reversals (marked by the vertical lines). The magnetic field strength displays an "unsettled behavior" during the years 2000–2002, with several short-duration reversals. In order to see how the HL CMEs are related to the polarity reversals in the individual hemispheres, we have separated the HL CME rate into northern (NHL) and southern (SHL) components (see Figure 3, middle panel [in the original article]). We have also indicated the 3 σ (standard deviation) rates by horizontal lines (*solid*: north; *dotted*: south) to assess the significance of the peaks. First of all, we note that there was a rapid increase in HL CMEs in the middle of the year 1999, especially in the northern hemisphere. After a local minimum, the NHL CME rate again had a broad maximum before dropping to a low value. On the basis of the magnetic field data, the north polar reversal was reported to be in 2001 February (Bilenko 2002) and 2001 May (Durrant & Wilson 2003). The north polar field strength was close to zero at these times. Figure 3 [in the original article] shows that the north polar field strength was close to zero at these times. In 2000 October there was a definite reversal (*thick line*), which coincided with the times of PCF disappearance (Lorenc, Pasorek, & Rybansky 2003; Harvey & Recely 2002; Gopalswamy et al. 2003a). The SHL CME activity picked up around the same time as the NHL CMEs, but the activity continued beyond the year 2002, subsiding after a large peak in the first half of 2002. The south polar reversal was reported to be in 2001 September (Durrant & Wilson 2003) and 2002 January (Bilenko

2002), consistent with the dashed vertical lines in Figure 3 [in the original article]. However, the times of PCF disappearance are much later: 2002 February (from Lorenc et al. 2003) and 2002 April (Harvey & Recely 2002). The cessation of HL prominence eruptions was in 2002 May (See Figure 1 [in the original article]). Note that the three prominence-related epochs coincide with the largest peak in the SHL CME rate. If we take a careful look at the magnetic field plot in Figure 3 [in the original article], we see that the south polar field was close to zero around this time before assuming a steady reversal. Thus the cessation of HL activity in the northern and southern hemispheres occurred in 2000 November and 2002 May, respectively, roughly marking the epochs of polarity reversal. The last SHL peak is likely to be due to the eruptions associated with the second tier PCFs in the southern hemisphere that reached latitudes exceeding 65° as evidenced by Ha synoptic charts (McIntosh 2003).

2.4. Comparison with Cycle 21 CME Rate

Although the observed CME rate for cycle 21 was a factor of 2 lower than the *SOHO* rate because of the poorer dynamic range of the earlier instruments, we were able to identify periods of HL CME cessation from both hemispheres. In Figure 4 [in the original article], we have shown the HL and LL CME rates for cycle 21 along with the NSO/Kitt Peak polar magnetic field data. Polarity reversal times from Webb et al. (1984) are also shown. The polar field was unsettled starting in the middle of 1980 all the way to the beginning of 1983, as shown by the magnetic field strength. There were corresponding

"temporary" reversals indicated by the arrows from various sources. Interestingly, there were also peaks in the HL CME rates from both poles approximately at the times of the these temporary reversals. The largest NHL peak for cycle 21 (toward the end of 1981) was bracketed by the disappearance of the northern PCF according to Webb et al. (1984) and Lorenc et al. (2003). The first SHL CME peak was close to the south reversal (1980 September) obtained by Webb et al. (1984), while the second peak was bracketed by the disappearances of the southern PCF found in Webb et al. (1984) and Lorenc et al. (2003). There is a large spike (more than 2 σ) in the NHL CME rate right at the time the north pole completely reversed around 1982 July. Note that the last period of north polar reversal listed by Webb et al. (1984) was 2002 February–March, a few months before the cessation of NHL CMEs indicated by the above NHL spike. In the south polar region, the CME rate was generally low, but there was indeed a small peak (1 σ) just before the time of complete reversal in the south. Despite the fact that the CME rate data were more noisy for cycle 21, we can clearly see that the polarity reversals are marked by the cessation of HL CMEs. Webb et al. (1984) noted that the PCFs disappear months after the magnetic polar reversal. However, this is true only when the first episode of reversal is considered. The final (or complete) reversal occurs only around the time of the cessation of HL CMEs (same as PCF disappearance). The first large spike in the LL CME rates in Figure 4 [in the original article] occurs around the times of the PCF disappearance. This may be due to active longitudes, similar to the large spikes in the LL rates for cycle 23.

3. Discussion

The results presented here bring out an important connection between the polar reversal as observed in the photospheric field and the coronal closed field as inferred from CMEs. This connection is strengthened by the fact that eruptive filaments are often found in the interiors of CMEs. We infer that the disappearance of PCFs, a traditional signature used for identifying polarity reversal, is a violent process involving CMEs of mass a few times 10^{15} g and a velocity of hundreds of kilometers per second. The kinetic energy of each of these CMEs is typically a few times 10^{30} ergs. The CME process helps the closed field lines overlying PCFs, and the filaments themselves become open to complete the polarity reversal. The results presented here also support the hypothesis of Low (1997) that CMEs may represent the process by which the old magnetic flux is removed and replaced by the flux of the new magnetic cycle.

Considering the HL CMEs also provides a natural explanation for the confusion regarding the actual epoch of polarity reversal. Wang, Sheeley, & Andrews (2002) compared the time of polarity reversal in cycle 23 from source surface predictions with the time of peak HL streamer brightness (2000 February) and found them to be roughly consistent for the north pole. However, they seem to have considered the first temporary reversal (see Figure 3, bottom panel [in the original article]). One would not expect the polarity reversal to coincide with the peak of HL streamer brightness, because the streamers would mean the presence of closed field structures near the pole (also consistent with the enhanced coronal brightness in the polar region as derived from coronal green line data; see Lorenc et al. 2003). The HL CMEs would result in the disappearance of these closed field structures before the polarity completely reverses. The polarity reversal indeed occurred toward the end of the year 2000 in the north polar region when the HL streamer brightness declined significantly (see their Figure 4). In the south polar region, the streamer brightness did not decline significantly until the end of 2001; this is consistent with the reversal in early 2002 as indicated in our Figure 3 [in the original article]. The decline in HL streamer brightness is therefore consistent with the involvement of HL CMEs in the polarity reversal process. The HL CMEs thus provide a mechanism by which the neutral lines that reach the poles, as indicated by several observations including the "coronal activity waves" (Benevolenskaya, Kosovichev, & Scherrer 2002), disappear, enabling the polarity reversal. The CME involvement also provides a more complete picture than the photospheric flux-cancellation mechanism.

4. Conclusions

The primary result of this Letter is that the epochs of solar polar reversal are closely related to the cessation of HL CME activities, including the nonsimultaneous reversal in the north and south poles. We have shown this to be true for solar cycles 21 and 23, for which complete CME data are available. Before the completion of the reversal, several temporary reversals take place with corresponding spikes in the HL CME rates. The HL CMEs also provide a natural explanation for the disappearance of closed field structures that approach the poles, which need to be removed before the reversal could be accomplished. Inclusion of CMEs along with the photospheric and subphotospheric processes completes the full set phenomena that need to be explained by any successful theory of the solar dynamo.

References

Benevolenskaya, E. E., Kosovichev, A. G., & Scherrer, P. H. 2002, in Solar Variability: From Core to Outer Frontiers, ed. A. Wilson (ESA SP-506; Noordwijk: ESA), 831

Bilenko, I. A. 2002, A&A, 396, 657

Brueckner, G. E., *et al.* 1995, Sol. Phys., 162, 357

Cliver, E. W., St. Cyr, O. C., Howard, R. A., & McIntosh, P. S. 1994, in Solar Coronal Structures, ed. V. Rusin, P. Heinzel, & J.-C. Vial (Bratislava: VEDA), 83

Durrant, C. J., & Wilson, P. R. 2003, Sol. Phys., 214, 23

Gopalswamy, N., Lara, A., Yashiro, S., Nunes, S., & Howard, R. A. 2003a, in Solar Variability as an Input to the Earth's Environment, ed. A. Wilson (ESA SP-535), in press

Gopalswamy, N., Shimojo, M., Lu, W., Yashiro, S., Shibasaki, K., & Howard, R. A. 2003b, ApJ, 586, 562

Gopalswamy, N., Yashiro, S., Nunes, S., & Howard, R. A. 2003c, Adv. Space Res., in press

Harvey, K., & Recely, F. 2002, Sol. Phys., 211, 31

Lorenc, M., Pasorek, L., & Rybansky, M. 2003, in Solar Variability as an Input to the Earth's Environment, ed. A. Wilson (ESA SP-535), in press

Low, B. C. 1997, Geophys. Monogr., 99, 39

Makarov, V. I., Tlatov, A. G., & Sivaraman, K. R. 2001, Sol. Phys., 202, 11

McIntosh, P. E. 2003, in Solar Variability as an Input to the Earth's Environment, ed. A. Wilson (ESA SP-535), in press

Wagner, W. J. 1986, in Coronal and Prominence Plasmas, ed. A. I. Poland (NASA CP-2442; Washington, DC: NASA), 215

Waldmeier, M. 1960, Zeitschrift fur Astrophys., 49, 176

Wang, Y.-M., Sheeley, N. R., & Andrews, M. D. 2002, J. Geophys. Res., 107, 10

Webb, D. F., Davis, J. M., & McIntosh, P. S. 1984, Sol. Phys., 92, 109

Many of the documents included in this anthology are research papers that were authored for peer-reviewed journals, or publications that are mostly read by professors, researchers, and other experts within the field covered by the particular journal. "Understanding and Predicting Space Weather" does not fall into that category. It is an article that was written for the general public. As such, whenever a concept like solar weather, magnetism, or solar wind is addressed, the author provides detailed explanations for these terms. The assumption is that the reader has no prior knowledge of the subject. For a peer-reviewed journal, the author or authors assume that readers are already familiar with subjects relating to the field that is covered by the journal.

Author Dawn Lenz provides a thorough overview of many of the processes previously mentioned in this anthology. She also links space weather to events here on Earth. Previously, most of the discussion has centered upon Sun happenings in interstellar space or near and within

Earth's magnetosphere and magnetopause. Now you will learn that solar weather can impact daily activity on our planet. Since solar storms consist of electromagnetic energy, they affect electromagnetic systems on Earth. As examples, the author mentions how space weather can change, and even damage, electric power grids, satellites, and other electricity-dependent systems. Since our reliance on such systems has grown over the years, our need to understand and predict solar weather also continues to grow. —JV

"Understanding and Predicting Space Weather"
by Dawn Lenz
The Industrial Physicist, December 2003/ January 2004

The consistency on Earth of visible solar radiation belies the sun's dynamic and turbulent state. Just beneath the solar surface, or photosphere, a layer of ionized hydrogen (along with a little helium and traces of heavier elements) churns and mixes to a depth of about 200,000 km, convecting heat from the 15-million-kelvin core to the 5,800-K surface. The churning charged particles generate electromagnetic fields that blossom from the sun's surface in spectacular patterns, which are observed in the tenuous, 1-million-kelvin plasma of the solar corona. The corona forms the base of the solar wind, the continuous, even-more-tenuous stream of charged particles that flows outward from the sun into interplanetary space. The effects of the interaction of solar charged particles with Earth's magnetic field are referred to as space weather.

Like terrestrial weather, space weather is characterized by an average state of relative calm punctuated by bursts of activity. These solar storms vary in strength and frequency with the 11-year solar-activity cycle and cause disruptions of various magnitudes on Earth. During calm periods, the only manifestation of solar weather may be the auroras (Northern or Southern Lights), caused by the excitation of atmospheric oxygen and nitrogen by the solar wind's energetic electrons. Flares and coronal mass ejections (CMEs) are two types of solar eruptions that can spew vast quantities of radiation and charged particles into space, potentially causing geomagnetic storms. If a large flux of charged particles from the sun intersects the Earth, it can have serious consequences for modern support systems, including electrical power grids, communications networks, and satellite operations.

Flares and CMEs differ spatially and temporally. Flares are strong transient outbursts of radiation, released near the solar surface, that extend tens or hundreds of thousands of kilometers into the outer solar atmosphere (Figure 1 [in the original article]). They are highly localized on the sun. Flares typically last for a few minutes to a few hours, and they emit radiation across most of the electromagnetic spectrum. Most of a flare's energy is released as radiation in the corona, but some energy contributes to forcing electrons and ions through the outer solar atmosphere and into the interplanetary solar wind.

CMEs are slower to develop (they emerge from the sun over the course of a few hours) and have spatial extents many times that of flares. Most of their energy is expended in driving ionized particles into interplanetary space rather than in radiation (Figure 2 [in the original article]). The angular size of a CME can range from a few degrees up to half a solar hemisphere or more. If a flare is analogous to an interplanetary thunderstorm, a CME initiates an interplanetary tsunami—a flood of billions of tons of protons and electrons bursting from the sun that is capable of massive interference with any flux-sensitive apparatus it happens to encounter.

Solar Storms on Earth

As technology advances, populations grow, and urban industrialized areas sprawl, Earth becomes more dependent upon systems that are vulnerable to damage from solar storms, including electrical grids and the swarm of satellites in orbit above Earth's protective atmosphere. Today's electrical grids are more susceptible to solar-storm disruption than their more localized predecessors because of the large geographical areas they cover and their interconnected nature. Communications systems and networks have developed beyond ground-based lines to satellite-based transmissions. Humans and their support systems venturing more extensively beyond the safety of Earth's atmosphere and into orbit, to the moon, or one day to the planets are largely unshielded from the solar storms that Earth's magnetosphere deflects at home (Figure 3 [in the original article]).

Satellite-based activities and operations are also vulnerable to the direct impact of a flux of solar energetic particles. About 150 satellites currently orbit Earth hundreds to thousands of kilometers above the top of the atmosphere for the purpose of relaying television and telephone signals at very high to ultrahigh frequencies (VHF/UHF). Both frequency ranges are used because their short wavelengths can penetrate Earth's ionosphere with minimal reflectance and interference. However, VHF and UHF wavelengths are not short enough to afford them complete immunity to atmospheric interaction, and they are susceptible to disruption from significant modulations in the ionosphere, which can occur during solar storms. One such storm occurred on July 14, 2000, when a large flare bombarded Earth with energetic particles that disrupted communications and associated support systems. Weather satellites returned pictures blurred by static, commercial fishing boats lost radio communication, and power companies in the northeastern United States had to reroute electricity in response to voltage disruptions.

In addition to operational interference, satellites and power grids can suffer physical damage from solar storms. Satellites draw power from solar cells, which consist of semiconductor materials that are sensitive to energetic ions. The continual flux of solar particles gradually degrades the effectiveness of solar cells, eventually crippling the satellites when the cells can no longer generate the required power. Solar storms significantly accelerate such degeneration. A

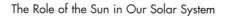

single strong solar storm can decrease the lifetime of a satellite's solar-cell system by several years.

Electromagnetic systems are vulnerable to electromagnetic-field fluctuations induced by a rapid influx of charged particles. Within power grids, geomagnetic storms can cause large-scale fluctuations and outages. Perhaps the most notorious solar-induced power outage occurred on March 13, 1989, in Quebec, when 6 million people experienced a 9-h electrical blackout caused by a CME. In addition to causing a loss of power, such events can damage power-grid hardware as abnormally large currents and voltages overload the system. Widespread power outages and communications breakdowns can cost millions of dollars; the 1989 Quebec outage cost an estimated $300 million. On a national scale, the economic impact of such an event can be in the billions of dollars.

Eruption Physics

Our observational picture of solar flares and CMEs has improved dramatically over the last decade with the inception of state-of-the-art solar telescopes and satellite-borne instruments, such as the Solar and Heliospheric Observatory, the Transition Region and Coronal Explorer, and, most recently, the Ramaty High Energy Solar Spectroscopic Imager. However, the detailed underlying physical causes of solar storms largely remain a mystery.

Theories of how flares and CMEs develop and erupt, the conditions in the solar atmosphere required for the generation of such phenomena, and the mechanisms by which the energy is expelled and the particle flux is propelled outward into interplanetary space are areas of active investigation in solar physics. The foundation of almost all such theories involves the twisting and tangling of magnetic-field lines in the solar atmosphere as a result of the underlying fluid motions in the convective layer just beneath the solar photosphere. According to the theory of magnetic reconnection, developed by Eugene Parker of the University of Chicago and Peter Sweet of the University of Glasgow (Scotland) in the 1950s, solar magnetic-field lines progressively become more chaotically intertwined, increasing the stresses between them (Figure 4 [in the original article]). When the stresses become severe enough, the field lines reconnect with an associated release of energy.

Flares and CMEs are sometimes observed to occur together. Until recently, this observation compelled researchers to look for a causal relationship between the two. Although both types of eruptions are believed to have physical roots in magnetic reconnection, solar physicists generally no longer envision a causal relationship and treat each separately in doing phenomenological modeling. Similarly, solar physicists believed for decades that flares caused geomagnetic storms. Such a correlation seemed plausible, given the enormous energy fluxes observed in flares, and solar-terrestrial storms do sometimes appear to be correlated with solar flares. However, explaining the physical correlation proved to be a challenge because there are both temporal and spatial inconsistencies between flares and geomagnetic storms. Flares typically last for at most a few hours and are highly localized. Storms can last for days and cover many times the area of flares.

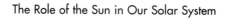

The key player in major solar-terrestrial events is now thought to be the CME rather than the flare. CMEs went unrecognized as significant solar phenomena for many decades after flares first received close attention, in part because CMEs produce less radiation than flares and require more sensitive and careful observation. Rather than expelling energy predominantly in the form of radiation and localized particle acceleration, a CME uses its energy to propel ions and electrons into interplanetary space.

Currently, the generally accepted model of the largest solar-terrestrial events is that they are caused by the acceleration of interplanetary charged particles ahead of a CME-induced shock. The triggers of CMEs, however, remain under debate as scientists pursue observational data to test various theories. Two competing views are (1) CMEs are triggered by the twisting and subsequent reconnecting of magnetic-flux ropes beneath the solar surface, with the released energy forcing particles out from inside the sun, and (2) CMEs, like flares, are triggered by the release of magnetic energy in the corona, above the solar surface (Figure 5 [in the original article]).

Predicting Storms

The peak of the last 11-year solar cycle, with a corresponding peak in flare and CME events, was in 2000, when Earth was significantly more dependent on power grids and satellite-based communication than during the previous peak. This dependency, coupled with new knowledge about the causes and effects of solar storms, spurred efforts to predict large geomagnetic storms in hopes of mitigating their effects.

As in meteorology, the tools of space weather forecasting include observations and model predictions. Observational data include in situ measurements of radiation and energetic particles at satellite orbit altitudes, and ground-based magnetometer data. In addition, solar-physics research satellites can provide data on current conditions at the sun. However, their instruments collect high-resolution data of just a few percent of the solar disk at a time, so only events occurring in the field of view for a specific observation sequence are captured. Space-weather modeling aims to take observational data as input and help forecasters predict storms. This relatively new field has grown significantly in recent years; about 70 % of the existing academic literature on space-weather modeling has been published since 2000. As interest in space-weather prediction increases, the models continue to improve.

Satellites designed for space-weather exploration include Wind (launched in 1994), the Advanced Composition Explorer (1997), and the Imager for Magnetopause-to-Aurora Global Exploration (2000). Their instruments gather radiation and particulate data to discover the characteristics of the interaction between the solar wind, solar energetic particles, and Earth's magnetosphere. On Earth, networks of ground-based magnetometers detect fluctuations in the planet's magnetic field. Solar storms commonly induce fluctuations on the order of 1 % in the measured magnetic field; magnetometers can detect fluctuations several

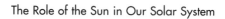

orders of magnitude smaller. Together, satellite and magnetometer data can provide accurate, up-to-the-minute space weather forecasting.

The first commercial space-weather prediction system was installed in England in January 2000. SpaceCast/PowerCast, developed by the Metatech Corp. (Goleta, CA), collects up-to-the-minute data from a group of satellites and networks of ground-based magnetometers about the sun's radiation and magnetic-activity levels. Predictive modeling is coupled with observational data to create specific regional forecasts. The system provides advance warning of an impending solar storm, permitting crucial or sensitive power-grid components to be shut down or otherwise protected. Devices that block anomalous currents are expensive to install on a large scale, however, so disabling essential components is currently the most cost-effective way to prevent damage.

As with severe terrestrial storms, the effects of solar storms can be mitigated with accurate and expeditious forecasting. The ability to predict major solar storms can give power companies sufficient lead time to implement preventive measures. Like sandbagging and nailing boards over windows before a hurricane, contingency strategies cannot disarm a major geomagnetic event, but they can significantly lessen its impact. Advance warning of storms can also, in principle, allow communications companies to notify their customers that a lapse in service may be imminent and estimate how long the lapse might last.

Our understanding of both the causes and the terrestrial effects of space weather is a subject of active research. Industrial focus on geomagnetic storms has thus far been motivated by efforts to reduce their impact, but just as we have learned to capture solar radiation and wind energy for modern power applications, we may one day learn to lasso and exploit the energy that reaches us in solar storms.

Reprinted with permission from "Understanding and Predicting Space Weather," by Lenz, Dawn, *The Industrial Physicist*, December 2003/January 2004, pp. 18–21. © 2004, American Institute of Physics.

The prior article mentions how solar storms have impacted activities on Earth, but "Stormy Weather: When the Sun's Fury Maxes Out, Earth May Take a Hit" focuses on those events with a bit more dramatic detail. On July 14, 2000, the Sun had just come to the end of its eleven-year cycle of polar reversal. A lot can happen in eleven years. For example, in 1989, not all households, businesses, and schools had more than one computer, if any computer at all. The 1990s saw an explosion of computer use and the further evolution of the Internet. By 2000, most of us had become heavily dependent on computers and their sometimes delicate inner electrical workings.

When the July solar storm hit that year, electrical systems on Earth received a jolt. While few outside of the scientific community noticed the storm, it serves as a warning for what could happen in the future when solar energy explodes into its eleven-year rage. Just above Earth on that July

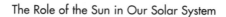
*day, havoc ensued. A number of satellites were
damaged, and the storm shut down East Coast
voltage regulators. Author Ron Cowen mentions
the damaging force of "killer electrons," or particles emitted by the Sun that each have energies
that can exceed 1 million electron volts, which is
an almost unthinkable amount of energy. —JV*

"Stormy Weather: When the Sun's Fury Maxes Out, Earth May Take a Hit"
by Ron Cowen
Science News, January 13, 2001

On July 14, 1789, an angry crowd stormed the infamous Bastille prison. This act of defiance ignited a
revolution that turned the streets of Paris crimson, as
mobs carried aloft the guillotined heads of aristocrats.

This past July 14 also marked a time of horrific violence, but on a much larger scale.

At 5:03 a.m. eastern time, a region on the sun large
enough to swallow Earth suddenly became 10 times
brighter, firing a torrent of high-energy radiation into
space. Then, the sun's outer atmosphere belched a billion-
ton cloud of magnetized, charged particles. Traveling
more than 6 million kilometers an hour, the magnetic
cloud headed straight for Earth.

Twenty-five hours later, the cloud hurtled past the
Solar and Heliospheric Observatory (SOHO), a European
Space Agency-NASA satellite that continuously monitors
the sun. One of the craft's solar panels—its power
source—suffered in seconds as much damage as it

normally accrues in a year's exposure to the harsh environment of space. Another craft, NASA's flagship X-ray observatory, Chandra, was forced into hibernation.

The cloud also temporarily knocked out the sun sensor on a satellite that measures the solar wind, the breeze of charged particles that blows out from the sun. Operating blind, the craft didn't know where to point for several days.

ASCA, a Japanese X-ray satellite, was even less lucky. Accelerated by the cloud, charged particles fried the craft's flight computer, spinning ASCA out of control. The craft never regained power. Reduced to a frozen piece of space junk, ASCA will later this year crash into Earth's atmosphere.

About 26 hours after it shot out from the sun, the cloud reached Earth. Ramming into the magnetosphere, the invisible magnetic shield that surrounds our planet, the storm revved up charged particles and dumped the equivalent of 1,500 gigawatts of power into the atmosphere. That's four times the power generated by the U.S. power grid. The disturbance severely damaged two large power transformers and disturbed electric power systems, shutting down voltage-regulating devices all along the East Coast.

Scientists forecast storms as severe, or even worse, over the next 2 years.

Welcome to solar maximum.

Pole Reversals

Like Earth's magnetic field, the magnetic field of the sun has a north pole and a south pole. But every

11 years, those poles reverse direction—with great commotion. The peak of that activity is called the solar maximum, which scientists determine by counting the dark blotches on the sun, where bundles of magnetic field lines concentrate. The number of these sunspots appears to have peaked last summer, more than 6 months later than researchers had predicted. The peak, however, is a broad one, and the solar maximum is expected to last for another 18 months.

Until the Bastille Day event, this solar maximum—the 23rd on record—seemed relatively puny. Even now, after solar storms pummeled Earth twice last November, it isn't likely to stand out as one of the strongest, says forecaster Ernest Hildner, director of the National Oceanic and Atmospheric Administration's Space Environment Center in Boulder, Colo. But in two other respects, he notes, this solar cycle is like no other.

With 2,000 communications satellites launched since the last solar maximum, astronauts taking longer trips in space, and a society ever more dependent on computers, cell phones, automated banking systems, and other electronic equipment, Earth has never been as vulnerable to the havoc that an electric disturbance from the sun can wreak.

In May 1998, for instance, a communications satellite called Galaxy IV abruptly failed. Although the cause of the failure is not known definitively, it occurred just after an unusually intense period of solar activity. When the satellite went belly-up, 45 million pagers suddenly went dead.

Characteristics of the power grids that transmit electricity make matters worse, says power-systems engineer John Kappenman of Metatech Corp in Duluth, Minn., a firm that monitors solar disturbances for several businesses. Today, power is being transmitted over greater and greater distances to more and more customers, yet utilities haven't proportionately increased the number of devices they employ to regulate voltage, he notes. Because they help minimize any sudden surge in current, such devices play a critical role when a solar storm hits Earth.

"In general, power grids are more vulnerable than they were 10 years ago," Kappenman says.

At the same time, never before have planetary scientists had so large a flotilla of spacecraft to monitor the sun and its effects on Earth. SOHO, NASA's TRACE (Transition Region and Coronal Explorer), and the Japanese satellite Yohkoh track aspects of the sun's roiling physics. Also, the European Space Agency's Ulysses craft is getting ready to make its second tour of the sun's poles.

But wait, there's more. ACE (Advanced Composition Explorer), like SOHO, lies 1 million miles closer to the sun than Earth does. By monitoring changes in the space environment at that location, it provides warning of a solar storm 1 hour before it reaches our planet. There's also WIND, which measures the solar wind that fills interplanetary space not far from Earth. Depending on its density and speed, this wind can boost or diminish the force of a solar eruption as it impinges on planets.

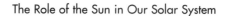

Another spacecraft, called IMAGE (Imager for Magnetopause-to-Auroral Global Exploration), is the first to observe charged particles and neutral atoms throughout the magnetosphere. Furthermore, both IMAGE and the Polar spacecraft examine auroras, the shimmering electric disturbances at Earth's poles generated when charged particles from the sun crash into the atmosphere. Meanwhile, Japan's Geotail spacecraft examines the back side of Earth's magnetosphere.

"These spacecraft are a wonderful way to understand cause and effect," says NOAA's Hildner. "You can watch an event leave the sun and come all the way to Earth, and you can measure it with this constellation of spacecraft. This is how we develop our models."

Solar Eruptions

Solar eruptions generally fall into two classes: flares and coronal mass ejections (CMEs). Both are generated by the sudden release of energy stored in a magnetic field, although scientists are unsure exactly how.

Magnetic field lines emerging from deep within the sun form giant, arching loops in the sun's corona. As the sun rotates, these loops become twisted, tangled, or stretched, storing vast amounts of energy. When stretched too far, the loops can suddenly snap or rearrange themselves, generating the largest explosions in the solar system. Some of the energy is released as a burst of radiation—a flare.

When the ultraviolet and visible light from a flare heads in our direction, it takes just 8 minutes to reach Earth. The greatest danger comes from the solar protons that may be accelerated by this explosive release of radiation, which arrive about 20 minutes after the flare does. If high-energy protons happen to strike astronauts outside the shelter of their spacecraft, they could be severely injured or even kill them. Because the ionosphere absorbs much of the protons' energy, they don't pose a threat to people or electrical systems on Earth.

There's a second way that a twisted magnetic field may release its energy. Like a stretched rubber band shot across a room, the field may fling itself from the sun's outer atmosphere, or corona, carrying the ionized gas surrounding it. These are CMEs—billion-ton parcels of ionized gas, or plasma, and the magnetic field holding them together. Sometimes referred to as magnetic clouds, these parcels can be bigger than planets and have much greater impact on Earth than flares.

Like a plane racing through the air at supersonic speed, a CME hurtles through the solar wind, creating a shock wave that accelerates the charged particles it meets. CMEs shoot radially outward from the sun, and only a few are directed toward Earth. When a CME and its charged-particle entourage strike Earth's magnetosphere—usually 3 to 4 days after the CME erupts in the corona—the cloud can trigger potentially devastating electrical events.

So-called killer electrons, which are charged particles revved up to energies greater than 1 million electron-volts, punch through the skin of spacecraft. If enough charge builds up in delicate electronic components inside the craft, electric arcing occurs. Jumping from one piece of electronics to the next, these tiny bolts of electricity

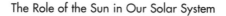

can destroy a satellite's sensitive electronic equipment, including computers and communications devices.

By compressing and jostling Earth's magnetosphere, the shock wave barreling in front of a CME delivers a punch that energizes charged particles within this magnetic shield. The magnetosphere may suddenly shrink, as it did in the aftermath of the Bastille Day storm. Normally extending 64,000 km from Earth's surface, the magnetosphere constricted to nearly half that length, notes space scientist Nicola J. Fox of NASA's Goddard Space Flight Center in Greenbelt, Md. She and other scientists described the Bastille Day eruptions late last month at a NASA press briefing.

Some CMEs are more dangerous than others. If one reaches Earth when its magnetic field happens to point southward, opposite to the direction of Earth's magnetic field, then the cloud can directly connect with the terrestrial field.

A southward-pointing CME dramatically alters Earth's magnetic field. In so doing, it creates a strong electric current, or ring current, which partially girdles the magnetosphere's equator and is much greater on Earth's nightside than dayside.

By linking to currents much closer to Earth, the ring current plays a pivotal role in disturbing power grids on the ground. The ring current on the nightside completes a closed circuit by connecting with currents that flow along Earth's magnetic field into and out of the ionosphere, the electrically conducting layer of ionized gas in the upper atmosphere.

Within the ionosphere, the current is known as the electrojet, which is associated with Earth's auroras and normally circulates at polar latitudes. A surge in the ring current jolts the electrojet and may also push it to lower latitudes. In the northern hemisphere, this brings it—and the aurora—farther south, where the magnetic disturbances it creates on the Earth's surface can induce currents in transmission lines over highly populated areas and disrupt many power systems.

For several hours, until the CME passes by, Earth is at its mercy. Many of the disruptive effects of the July 14 eruption stem from its southward-pointing CME.

Another southward-pointing CME, which struck Earth during the last solar cycle, had far more serious consequences. In the wee hours of March 13, 1989, the Hydro-Quebec power company was operating normally. Then, Earth's magnetic field fluctuated violently in the region along the U.S.-Canadian border. Voltage sagged, and within 90 seconds, all of Quebec province went black. Six million people woke that chilly late-winter morning without heat or electric lights.

Earthling Predictions

So, how are earthlings doing in predicting when a solar storm will occur and how severe it will be? "If the weather predictions were as bad as our ability to predict [the onset of a solar storm], we'd fire all the weathermen," says solar physicist Craig DeForest of the Southwest Research Institute in Boulder, Colo.

But once a storm erupts "we're doing fairly well," in determining when it will strike Earth, argues Kappenman. He and other scientists credit the ACE satellite, which experiences the brunt of a solar storm an hour before Earth does, with warning satellite and power companies about the severity of approaching storms. "For the first time, we can do continuous predictive forecasts of geomagnetic storms," he says.

An hour's lead time is enough to batten down the hatches of a satellite or alert a power company that may need to rely on reserves and minimize use of long-distance transmission lines.

Scientists still have a long way to go before they can make reliable predictions days to weeks in advance, but there's been some progress. After reviewing more than 2 years of data on solar storms, researchers reported in 1999 that magnetically active regions on the sun often exhibit an S-shaped pattern in their X-ray emissions.

S marks the spot because the twisted or stretched magnetic field lines create this sinuous pattern. But researchers don't know how soon after an S appears an active region will erupt. And since the sun rotates once every 28 days, the timing is critical for determining if the region will be in line with Earth when it sends out a CME. "If you see an S shape, you know the gun is loaded, but you don't know when the gun is going off, so you don't know which way it will be pointed," explains DeForest.

Once an eruption occurs, an instrument on SOHO can determine whether or not the storm is headed Earthward. That gives forecasters 2 to 3 days, warning that something may hit, although it provides little information on severity.

Two independent teams of astronomers recently found different ways to discern activity on the sun's far side, its hidden half, days before it rotates into view. One technique relies on the detection of ultraviolet radiation emitted by hydrogen gas in the sun's vicinity. When energy from a solar outburst hits the gas, it creates ultraviolet hot spots that SOHO can detect, even if they occur on the sun's far side.

In the other method, researchers use an instrument on SOHO to listen in to the sun's acoustic vibrations. A slight increase in the velocity of sound waves bouncing off the sun's far side may alert scientists about a region likely to erupt before the far side rotates into view.

In another diagnostic effort, solar physicists have developed what appears to be a reliable method for predicting how long it will take a given CME to reach Earth. They determine the arrival time of a CME by taking into account how much the solar wind can speed up or slow down the storm.

Another team, which includes participants from the Air Force, is going after the greatest challenge in the space weather business—attempting to determine if the magnetic field of a CME will point south. In trying to model this critical feature, researchers must take into account not only the initial characteristics of a CME but also the speed and density of the solar wind through which it travels. The wind can reorient the CME's magnetic field into more or less dangerous directions, says Murray Dryer of the Space Environment Center.

Looking toward the future, two NASA craft set for launch in 2004 are intended to view the sun's corona from different directions, providing a three-dimensional view of solar storms. The mission, dubbed Stereo, is expected to identify those storms heading earthward more quickly than the current array of satellites does.

In 2006, the space agency plans to launch the Solar Dynamics Observatory as the next-generation SOHO. The craft will provide nearly 24-hour coverage of the sun, says project scientist Barbara J. Thompson of Goddard.

Researchers still hope to revive the concept of Geostorm, a proposed mission that was not funded this fiscal year. The craft would use ultrathin sails that take advantage of the push of sunlight to reach orbits closer to the sun than those of ACE. Geostorm could provide data on an impending storm 2 to 3 hours before it reaches Earth.

Such a mission probably wouldn't be launched until the next solar cycle—just in time to greet the next great wave of violence from the sun.

References

2000. NASA, NOAA gain unprecedented view of angry solar cycle. NASA press release. December 21. Available at ftp://ftp.hq.nasa.gov/pub/pao/pressrel/2000/00-199.txt.

Damage to satellites sounds like a remote threat to activities on Earth, but it is more direct than you might realize. As the article "Solar

Flare Could Disrupt Technology" points out, individual telephones, pagers, radio signals, satellite television services, and other systems that rely upon wireless signals are vulnerable to solar flares and other solar storm events.

Imagine that you are the designer of a new satellite radio service. You design your system to work effectively on a daily basis and you foresee what problems might occur to disrupt your service. You then build solutions into your system for when the worst happens. But what about threats that are not fully understood? In many ways, preparing for a solar storm attack is like preparing for a terrorist attack. The general information is understood, but the details often remain fuzzy. That is why organizations like the National Oceanic and Atmospheric Administration (NOAA) continue to study solar flares to understand how they can interfere with technology-based systems. —JV

"Solar Flare Could Disrupt Technology"
by Cade Metz
PC Magazine, October 24, 2003

Set in motion by an eruption of gas on the sun, an enormous space storm—known as a coronal mass ejection, or CME—was headed towards earth Friday afternoon, and it could cause problems with satellites, cell phones, pagers, and other technological equipment. The storm has already interfered with high-frequency airline communications and power grids.

Earlier this week, forecasters at the National Oceanic and Atmospheric Administration's Space Environment Center in Boulder, Colorado, noticed two enormous sunspots, one that eventually grew to about ten times the size of the Earth. It is unusual to see such spots at this point in the sun's cycle.

"It's like seeing a hurricane in November rather than August, when you'd typically expect it," says Larry Combs, one of the Space Environment Center forecasters. "The peak of the cycle was in 2000, and here we are 3 and half years later with a dynamic sunspot region that you could put 10 earths inside."

Then, on Wednesday morning, at 3 am EDT, the larger of the two spots produced the CME, which is headed in the direction of earth, achieving speeds of 2 million miles per hour.

According to the NOAA, the sunspot already produced a major solar flare earlier in the week, causing a radio blackout on October 19 at 12:50 pm EDT. Today's CME could cause similar problems.

"It probably is already having an effect on satellites," says Combs. "This is a moderate to strong magnetic storm. With the storm at its current altitude and intensity, satellites will experience some surface charging, which causes problems with satellite components as well as satellite drag, where their orbits are affected."

This could have an impact on phone and pager networks that tie into satellites as well as satellite-based radio and Internet services. Satellites aside, the storm could directly affect individual telephone, pager, radio signals, and other wireless signals. "But if it does, problems will only be periodic," says Combs.

Electrical utilities are also at risk. "The power grids are already seeing effects," says Combs. "Some voltage corrections will be needed, but power companies know how to deal with these things." And, as the Associated Press reports, the storm is already interfering with airlines communications. "Planes are operating up in the atmosphere, and they are experiencing radio blackouts, especially at the high latitudes," continues Combs.

Representatives at both AT&T Wireless and Verizon Wireless said they don't expect any degradation of service, because the cell phone towers are located on the ground. A spokesperson at Cingular echoed this and added, "According to Cingular Wireless' radio transmission engineers, Cingular's GSM/GPRS, TDMA and Mobitex networks are terrestrial radio networks operating in the 800, 900, and 1900 MHz bands. Most solar radiation is absorbed by the Earth's atmosphere and does not affect terrestrial radio transmissions in these bands."

In recent years, scientists have made significant progress in developing ways to detect solar flares and coronal mass ejections that might break through Earth's magnetosphere to affect activities on our planet. One of these detectors is called the Solar Mass Ejection Imager (SMEI).

It is the subject of "Storm Spotting: A Step Closer to Forecasting Disruptive Solar Activity."

As you read this paragraph, the SMEI is orbiting Earth in search of solar storm activity. According to the article, the SMEI has detected seventy coronal mass ejections since its launch in January 2003. That is an impressive start. The successes belie a few important problems, however. As author Krista West points out, the SMEI could be missing some coronal mass ejections due to problems in distinguishing the solar storm output from other background light. Even worse, it takes approximately one full day for data from the SMEI to reach Earth. One can imagine that more time should be added to that figure, given the fact that scientists would have to process the data and release it to the public, if disruptive solar activity was headed our way. Despite the glitches, SMEI sounds like a promising solution to the problem of unpredictable solar storms. —JV

"Storm Spotting: A Step Closer to Forecasting Disruptive Solar Activity"
by Krista West
Scientific American, March 15, 2004

On October 19, 2003, a large solar flare erupted from the surface of the sun, drawing scientists' attention to three massive sunspot groups that, over the next two weeks, produced a total of 124 flares. Three of them were the biggest flares ever recorded. Along with these bursts of electromagnetic radiation came enormous clouds of plasma mixed with magnetic fields. Known as coronal mass ejections (CMEs), these unpredictable clouds consist of billions of tons of energetic protons and electrons. When directed earthward, CMEs can create problems. At last count, the fall's flares and CMEs affected more than 20 satellites and spacecraft (not including classified military instruments), prompted the Federal Aviation Administration to issue a first-ever alert of excessive radiation exposure for air travelers, and temporarily knocked out power grids in Sweden.

Historically, CMEs have struck the earth with little or vague warning. If they could be forecast accurately, like tomorrow's weather, then agencies would have time to prepare expensive instruments in orbit and on the ground for the correct size and moment of impact. Such precise predictions could soon emerge: last December researchers announced the early success of a forecasting instrument, called the Solar Mass Ejection Imager (SMEI), that can track CMEs through space and time.

Launched in January 2003 on a three-year test run, SMEI (affectionately known as "schmee") orbits the planet over the poles, along the earth's terminator, once every 101 minutes. On each orbit, three cameras capture images that, when pieced together, provide a view of the entire sky with the sun in the middle. The scattering plasma electrons of CMEs appear on SMEI images as bright clouds.

Other sun-watching instruments can image CMEs, but they work like still cameras, taking single pictures of the sun. NASA's Solar and Heliospheric Observatory

(SOHO), for example, can "see" CMEs erupting from the sun quickly but is soon blind to the path of the clouds. SOHO came in handy last fall when it caught two large CMEs headed for the earth, but it could not follow the ejecta nor provide an accurate impact time.

Instead of a SOHO-style snapshot camera, SMEI works more like a 24-hour surveillance system, constantly scanning and tracking. SMEI begins looking about 18 to 20 degrees from the sun and continues imaging beyond the earth. SMEI can determine the speed, path and size of a CME, allowing for refined and reliable impact forecasts. Such information is particularly useful, scientists say, in predicting small CME events. Such ejections can take anywhere from one to five days to reach our planet. Since its launch, SMEI has detected about 70 CMEs.

During last fall's solar storms, SMEI had its first big chance to prove worthy of its estimated $10-million price tag. Managed primarily by the Air Force Research Laboratory at Hanscom Air Force Base in Massachusetts, about 20 air force and university scientists have been developing SMEI over the past 20 years. At the December 2003 American Geophysical Union meeting in San Francisco, Janet Johnston, SMEI's program manager, proudly announced that SMEI had successfully detected two of the autumn's largest CMEs about 21 and 10 hours, respectively, before they struck the earth.

Unfortunately, scientists didn't know of the detection and tracking potential until after the storms hit the earth. Right now it takes about 24 hours for SMEI data to reach Hanscom because they travel through multiple ground-tracking stations. According to David F. Webb, a physicist at Boston College who is part of the SMEI team, precise forecasting demands a reduction in data-transmitting time from 24 to six hours. Such a reduction will require more researchers at ground-tracking stations to move information along and to inspect SMEI's output.

SMEI's data gathering may also need perfecting. Lead forecaster Christopher Balch of the Space Environment Center in Boulder, Colo., emphasizes that the CME signal must stand out better against other background light. Once improved, SMEI "could potentially fill a gap in our observations," Balch says, by allowing scientists to track CMEs precisely, thereby making "real-time" forecasts possible.

Reprinted with permission from Krista West.

4

The Sun and Earth's Climate

Previous articles in this anthology reveal how solar activity through flares, coronal mass ejections, and other solar storms affect electromagnetic systems near and on Earth. But can solar storms influence our weather? Already we have learned that solar weather and weather on our planet are two different processes. Even so, the Sun affects everything on Earth. Without the Sun, there would be no life on our planet. Weather, therefore, is not immune to the ever-changing forces of the Sun.

As the article "Earth's Response to a Variable Sun" indicates, recent studies show that solar changes do influence our global climate. This issue is a very politicized one. The basic question concerns whether human activities, such as burning fossil fuels, make Earth more vulnerable to solar variability, or solar forcing. Authors Judith Lean and David Rind do not address this issue but instead focus attention on the known science behind the Sun's ability to affect Earth's climate. It appears that during *peak phases of the Sun's eleven-year cycle, changes can be felt here in terms of daily weather and overall climate. —JV*

"Earth's Response to a Variable Sun"
by Judith Lean and David Rind
Science, April 13, 2001

"Since we are the children of the Sun, and our bodies a product of its rays . . . it is a worthy problem to learn how things earthly depend on this material ruler of our days." Thus wrote Samuel Pierpont Langley in *The New Astronomy* in 1898. But the "worthy problem" of how solar variability may affect processes on Earth remains unresolved after more than a century, despite the fact that it is of particular importance today because of its implications for global climate change. Even a modest influence of solar variability on Earth's climate alters the assessment of anthropogenic effects and their likely future impacts.

Recent investigations of Earth's surface, upper ocean, and lower tropospheric temperatures and the sun's irradiance suggest that there is indeed a discernable influence of solar variability on global climate. This follows a decade in which connections between solar variability and climate were alternatively pronounced and dismissed (1–3). But how solar variability is translated into climactic changes on Earth remains to be fully explained.

Solar irradiance varies slightly over an 11-year cycle. This cycle of the sun's magnetic activity alters its energy output, as well as the occurrence of sunspots, flares, and coronal mass ejections. Sunspots (see the dark regions on the solar disk in the [first] figure [in the original article]) have been used to track fluctuations in the strength of the sun's 11-year activity cycle for more than 300 years. Variations that are approximately in phase with this solar cycle have now been detected in satellite records of global temperatures in the lower troposphere as monitored by the Microwave Sounding Unit (MSU) since 1978 (4), in upper ocean temperatures since 1955 (5), and in surface temperatures, primarily from thermometer records during the past century (6). In recent decades, these 11-year temperature cycles had peak-to-peak amplitudes of 0.06¡ to 0.1¡C. They coincided with directly observed total solar irradiance changes $\triangle I$ of 1.1 W m-2 (0.08%) (7), which resulted in a climate forcing $\triangle F$ of 0.2 W m-2, where $\triangle F = 0.7 \triangle I/4$. For comparison, net climate forcing by anthropogenic sources is currently about 0.35 W m-2 per decade (8). The solar cycle signal becomes increasingly apparent in the upper troposphere and stratosphere, where it persists in four decades of global temperature records with amplitudes that increase from 0.3¡C at an altitude of 6 km to 0.9¡C at 20 km (9).

Decadal-scale climate oscillations by themselves do not prove a solar connection. Such variations have been linked to other phenomena, particularly ocean circulation (10, 11). But confidence in the reality of the sun-climate connection has grown as quantitative relations are established between increasingly long and precise databases of global temperature and solar irradiance, on decadal and centennial time scales, in combination with sophisticated statistical detection techniques and focused climate modeling studies.

In the 22-year-long, space-based MSU global temperature time series, a solar cycle-related periodicity appears after careful removal of the much larger (by a factor of five) influences of the El Niño Southern Oscillation (ENSO) and the El Chichon and Pinatubo volcanic eruptions (4). Analogous cycles were detected in an ensemble of century-long surface temperature data sets when signal processing was used to simultaneously identify the fingerprints of solar, volcanic, and anthropogenic influences (6).

Observations of total solar irradiance made with high-precision space-based radiometers since 1978 have provided quantitative measurements of solar forcing during the MSU epoch and, in combination with historical proxies of solar activity, have facilitated the reconstruction of longer term forcing. Paleoclimatic reconstructions of surface temperature and solar irradiance, although less certain than direct observations, augment and strengthen the case for sun-climate relations. It has been speculated (12) that solar irradiance changes by a few tenths of a percent on centennial time scales, with resultant climate forcing of 0.5 to 0.7 W m-2 and that these changes are associated with fluctuations of about 0.2¡ to 0.4¡C in preindustrial surface temperatures reconstructed from tree rings, ice cores, and corals (13). The empirical relation of "Little Ice Age" cooling and the lack of sunspots from around 1645 to 1715 (the "Maunder Minimum") is a recent example of such an association.

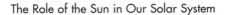

Various climate-modeling studies have used these temperature and irradiance results to explore the response of the climate system to direct and indirect solar forcing. Direct forcing in the troposphere occurs through solar changes primarily in the visible and near infrared spectrum (13, 14). A primary indirect forcing results from solar ultraviolet irradiance variations, which affect stratospheric ozone (15) and the mean winds. Ozone changes have a direct radiative impact on the troposphere, and altered mean winds affect planetary wave propagation from the troposphere and hence the natural modes of variability (16, 17). Observations have shown that descending large-scale circulation anomalies in the stratosphere influence tropospheric weather patterns (18). The fact that such mechanisms are seen in observations and have also been simulated in modeling studies of other forcings (such as CO_2 increase and volcanic aerosol injections) makes the solar variability influence less remarkable and more understandable.

The apparent link between Earth's temperature and variations in the sun's irradiance implies that climate responds directly to solar forcing with a sensitivity of 0.3¡ to 0.8¡C per W m-2 approximately in phase with the solar forcing. Equilibrium sensitivities from general circulation models (GCMs) are on the order of 0.7 . . . C per W m-2, so the magnitude of the solar variability signal in surface temperatures appears reasonable, at least on centennial time scales. But peak solar irradiance during a cycle lasts for only a few years. Accordingly, the climate is expected [from GCM studies such as (19)] to be less sensitive by a factor of 5 to 10 to forcing by the decadal solar cycle because ocean temperatures do not have time to fully respond. The apparent enhanced solar-climate sensitivity on decadal time scales thus requires further explanation, with the constraint that additional mechanisms seem not to be needed on the centennial time scale.

The increased temperature sensitivity at higher altitudes, where the amplitudes for the 11-year temperature cycles are larger than near Earth's surface, may indicate an atmospheric dynamical response to solar variability, operating through interaction with the stratosphere. Other more esoteric mechanisms include the potential effects of global cosmic rays and energetic particles (both of which vary with the solar cycle) on cloud condensation nuclei and cloud cover. Unfortunately, existing cloud climatology records are too short to test these ideas, and suitable observations of the response of free tropospheric aerosol to solar-induced ionization changes are also lacking.

Solar forcing may also amplify internal oscillations ("noise") in the climate system (such as ENSO) through processes such as stochastic resonance (20) or by exciting off-resonant responses in the delayed oscillator mechanism thought to be a component of many ENSOs (21). Given that ENSO modifies the climate of the continental United States (22), the long-standing empirical evidence linking U.S. drought to solar variability (23) may perhaps be explained by solar variability enhancement of natural climate variability modes in the Pacific. East African rainfall and drought reconstructions for the past 1,100 years also suggest that hydrological parameters may be especially responsive to solar variability (24), perhaps associated with an ENSO-like amplification.

Sun-climate connections are clearly complex. The superposition of direct and downward-propagating effects and enhanced internal variability are expected to produce complicated regional fingerprints. This complexity may account for the varying strengths of 11-year and other solar-type cycles (solar activity also exhibits cycles near 80 to 90 and 200 to 210 years, for example) in many climate records and the difficulty in identifying sun-climate connections.

A firmer understanding of the impact of solar variability on the climate system offers numerous potential benefits. By virtue of the multiple processes and feedbacks involved, sun-climate connections provide a unique test for numerical climate models, particularly their utility for predicting climate responses to future forcings, including anthropogenic effects. Understanding the linkages could facilitate use of the solar cycle in seasonal climate predictions if the magnitudes of the effect are found to be sufficiently large. Knowing how solar variability has altered climate in the past may help constrain the magnitude of anthropogenic warming and internal variability over the past century. It also provides a first-order indication of what may be expected because of solar variability in the future compared with other climate forcings (see the graph [in the original article]). If substantial greenhouse gas reductions are achieved, projected solar forcing may counter a substantial fraction of the remaining anthropogenic forcing in the next few decades, providing unexpected complications for climate change detection when increasing certainty is expected and needed.

References and Notes

1. R. A. Kerr, *Science* 254, 652 (1991).
2. ———, *Science* 268, 28 (1995).
3. ———, *Science* 271, 1360 (1996).
4. P. J. Michaels, P. C. Knappenberger, *Geophys. Res. Lett.* 27, 2905 (2000).
5. W. B. White, D. R. Cayan, J. Lean, *J. Geophys. Res.* 103, 21335 (1998).
6. G. R. North, Q. Wu, *J. Clim.*, in press.
7. C. Fröhlich, J. Lean, *Geophys. Res. Lett.* 25, 4377 (1998).
8. Intergovernmental Panel on Climate Change, *The Science of Climate Change*, J. T. Houghton *et al.*, Eds. (Cambridge Univ. Press, Cambridge, 1995).
9. H. Loon, D. J. Shea, *Geophys. Res Lett.* 27, 2965 (2000).
10. V. M. Mehta, T. Delworth, *J. Clim.* 8, 172 (1995).
11. G. R. Halliwell Jr., *J. Clim.* 10, 2405 (1998).
12. J. Lean, J. Beer, R. Bradley, *Geophys. Res. Lett.* 22, 3195 (1995).
13. T. J. Crowley, *Science* 289, 270 (2000).
14. D. Rind, J. Lean, R. Healy, *J. Geophys.* Res. 104, 1973 (1999).
15. J. P. McCormack *et al.*, *Geophys. Res. Lett.* 24, 2729 (1997).
16. D. Rind, N. K. Balachandran, *J. Clim.* 8, 2080 (1995).
17. D. Shindell *et al.*, *Science* 284, 305 (1999).
18. M. P. Baldwin, T. J. Dunkerton, *Nature*, in press.
19. J. Hansen *et al.*, in *Climate Processes and Climate Sensitivity*, J. Hansen, T. Takahashi, Eds., *Geophysical Monograph Series*, Vol. 29 (American Geophysical Union, Washington, DC, 1984), pp. 130–163.
20. A. Ruzmaikin, *Geophys. Res. Lett.* 26, 2255 (1999).
21. M. D. Dettinger, W. B. White, in preparation.
22. J. E. Cole, E. R. Cook, *Geophys. Res. Lett.* 25, 4529 (1998).
23. E. R. Cook, D. M. Meko, C. W. Stockton, *J. Clim.* 10, 1343 (1997).
24. D. Verschuren, K. R. Laird, B. F. Cumming, *Nature* 403, 410 (2000).
25. J. L. Lean, in preparation.
26. J. Hansen *et al.*, *Proc. Natl. Acad. Sci. U.S.A.* 97, 9875 (2000).
27. This work was supported by NASA.

Reprinted from SCIENCE. By Lean, Judith and David Rind, "Earth's Response to a Variable Sun," SCIENCE 292: 234-236 (2001).

"Pinning Down the Sun-Climate Connection: Solar Influence Extends Beyond Warm, Sunny Days" grapples with some of the problematic issues associated with the Sun and our current limitations in studying it. A good analogy is to think of weather forecasters who study daily natural fluctuations. Even in this age of high technology, weather forecasters still are not accurate 100 percent of the time. They may predict a sunny day for a day that turns out to be blustery. If weather forecasters cannot predict even local weather changes, you can imagine what a feat it is to try to understand and predict space weather events, much less to try and link these happenings to Earth's climate.

Notice in this article how some scientists argue that solar variations unaffected by human activities cannot account for the apparent global warming pattern we now are experiencing. That suggests other factors, such as human-produced pollution, could be to blame. Still other scientists argue that solar influence alone can explain the warming trend. The author of the article does a good job of balancing both sides of the highly politicized issue. —JV

"Pinning Down the Sun-Climate Connection: Solar Influence Extends Beyond Warm, Sunny Days"
by Sid Perkins
Science News, **January 20, 2001**

The solar flares that spew massive amounts of energy and particles earthward are notorious for the havoc they can wreak on satellites, power grids, and our planet's magnetic field. The charged particles that slam into the outer fringes of the atmosphere also ionize the air and stimulate shimmering auroras. During periods of increased solar activity, particularly during highpoints of the 11-year sunspot cycle, these breathtaking sky shows often appear far south of their normal Arctic venues.

Strong solar activity can also have substantial short-term influences by cooling the atmosphere in some places and heating it in others. These meteorological effects typically last days or weeks. But many scientists propose that changes in the sun's magnetic field and radiation output during its 11-year cycle of activity also have longer-term effects. They influence the movement of weather systems and other aspects of atmospheric and climate patterns.

It's tough to discern the subtle climactic effects of solar variation amidst a cacophony of strong earthly influences—greenhouse gases, volcanoes, sulfate aerosols, to name a few. But using sophisticated statistical analyses of an ever-growing stockpile of climate and weather data, scientists say they're uncovering ways in which even small variations in solar activity could have big effects down at ground level.

Solar Storms

Intense solar activity has substantial short-term effects that can shape Earth's weather. Consider the magnetosphere, which swaddles the planet at altitudes between 1,000 and 6,000 kilometers. It's composed of thick layers of charged particles—primarily protons and electrons—trapped in space by Earth's magnetic field. When a strong solar flare delivers a jolt to these belts, it can shake loose a torrent of these charged particles. Auroras, which at times of increased solar activity sprawl across the sky more often than average, are merely one result of this ionic rain upon the upper atmosphere.

An especially powerful type of solar eruption, known as a coronal mass ejection, can inject protons directly into Earth's atmosphere. As these high-energy solar protons rain down, they pummel the air like subatomic bullets. The kinetic energy of these charged particles may warm the outer layers of the polar atmosphere by several degrees, says Charles H. Jackman, an atmospheric scientist at Goddard Space Flight Center in Greenbelt, Md. This heating—which typically lasts just a few days, Jackman says—primarily affects the air at heights above 50 km.

At altitudes from 30 to 90 km, the energetic particles also split molecules of nitrogen into nitrogen atoms, which then react with oxygen in the air to form ozone-destroying compounds. This occurs mainly at polar latitudes, where most of the charged particles have spiraled in along lines of the Earth's magnetic field. Because ozone absorbs ultraviolet (UV) light, the depletion of ozone can cause the atmosphere to cool.

During a large solar storm, a portion of the upper stratosphere in the polar regions can lose up to 20 percent of its ozone and cool as much as 3°C, an effect that can last for several weeks. Jackman is quick to emphasize that such solar activity is not the cause of the infamous "ozone holes," which occur at lower altitudes and are aggravated by chemicals such as chlorofluorocarbons.

Driving Earth's Climate

Not only do solar storms influence day-to-day weather, but long-term, subtle variations in solar activity drive Earth's climate, scientists propose. Satellite data show that, overall, the amount of radiation Earth actually receives from the sun varies little between the maximum and minimum phases of the 11-year solar cycle.

When measured across all wavelengths of light, the total radiation reaching the planet changes only about 0.1 percent during the cycle, says Rodney A. Viereck, a space physicist at the National Oceanic and Atmospheric Administration in Boulder, Colo. However, the amount of change in solar irradiance at Earth isn't consistent across all wavelengths, Viereck notes. For example, the type of UV radiation that converts oxygen molecules into stratospheric ozone varies as much as 8 percent during the solar cycle.

At yet-shorter UV wavelengths—the ones that create the ionosphere, which is the layer of the atmosphere that lies above about 80 km in altitude—the disparity is even larger. During a solar maximum, Earth receives more than five times the amount of radiation in these far-UV wavelengths than it gets during a solar minimum.

The significant differences in UV irradiance at Earth during the solar cycle have profound effects in the upper levels of the atmosphere because the air density there is low, says Viereck. As these layers absorb radiation, they heat and expand upward, reaching high enough to slow satellites in low Earth orbit. Such spacecraft, which circle Earth at altitudes of 250 to 600 km, need to be boosted in their orbits more often during solar maxima than at other times, he adds.

Down at Earth's surface, the variations in solar irradiance during a solar cycle are relatively small. Many of the wavelengths whose intensities vary significantly are absorbed at higher altitudes, so they never reach the ground. Although no specific mechanism has been proposed to explain how solar activity might influence weather and climate, the solar cycle matches up with many earthbound patterns, says Judith Lean, a solar physicist at the Naval Research Laboratory in Washington, D.C. She recently reviewed data from a wide variety of research efforts. The list of phenomena following the 11-year cycle includes the frequency of rainfall in Africa, forest fires in North America, and hurricanes in the North Atlantic.

Another example of an apparent solar influence on climate is the 11-year cycle of annual temperatures in Alaska. Using Fairbanks and Anchorage weather data extending back to 1919, R. Suseela Reddy, a climatologist at Jackson (Miss.) State University, found that the annual average temperatures at the sites were lower during solar maxima and higher during solar minima. Reddy presented his research at the American Meteorological Society's annual meeting in Albuquerque this month.

At the same meeting, Alfred M. Powell Jr., an atmospheric scientist with Autometric in Springfield, Va., described a mechanism that he believes links changes in UV radiation that heat the upper atmosphere to effects on the weather at ground level.

According to Powell, as stratospheric ozone absorbs heat, the boundary between the stratosphere and the troposphere—the lowest atmospheric level, where all the weather happens—is driven to lower altitudes. In effect, Powell says, this clamps a lid on the weather systems in the troposphere. Low-pressure systems grow stronger and more quickly, and they last longer.

The lowered boundary also changes the circulation of air from the stratosphere downward into regions of low pressure, which, in turn, affects the flow patterns that steer weather systems. Previous research has shown that at solar maxima, weather systems are driven along a more southerly path and are stronger. Powell's model reproduces these characteristics.

Climate Changes

Only for the past two solar cycles have scientists used satellites to make direct measurements of the 11-year variation in radiation reaching Earth. It turns out, however, that these changes in irradiance correlate closely with the sunspot number, which astronomers have been measuring for hundreds of years. This correlation enables

climate researchers to look at sun-climate connections well into the past.

In particular, scientists are interested in explaining climate changes since the Industrial Revolution began to spew carbon dioxide and other so-called anthropogenic greenhouse gases into the atmosphere. A team led by Peter A. Stott, a climate scientist at the Hadley Center for Climate Prediction and Research in Berkshire, England, used a computer to model the changes in global average temperature since 1860. Greenhouse gases can't fully explain the climate patterns of the past century, the team reported in the Dec. 15, 2000 *Science*.

Even though carbon dioxide and other greenhouse gases have been rising constantly for the past 150 years, global average surface temperatures during that time have shown extended periods of both warming and cooling. After a period of warming early in the century, slow but steady global cooling began in the 1940s. It continued until the mid-1970s, when warming seems to have kicked in and overtaken the cooling trend.

Stott says that anthropogenic emissions alone could explain the rapid rise in temperatures in the past 30 years and that solar variation alone could have caused the warming observed between 1910 and 1940. However, neither natural nor anthropogenic causes alone could explain temperature patterns throughout the entire century. A model that includes both causes could explain up to 60 percent of the temperature fluctuations measured on land, Stott notes. The rest of the fluctuation can be attributed to normal year-to-year variations, he adds.

Although these results show that the solar variation can be a major contributor to climate change, Stott says he expects rising anthropogenic emissions to increasingly dominate as a cause of global warming.

Francis W. Zwiers, an atmospheric scientist at the Canadian Center for Climate Modeling and Analysis in Victoria, B.C., finds the British team's results compelling. In most runs of Stott's computer model, which uses a random statistical process in its calculations, changes in solar irradiance were needed to duplicate the century's temperature patterns, he says. Only in a few runs did natural year-to-year variation explain the climate changes.

Gerald R. North, a climatologist at Texas A&M University in College Station, contends that the effect of solar variation is so small that it's of no practical significance. This effect is "no more that a few hundredths of a degree . . . a faint signal in a noisy system," he says.

Teasing out the individual contributions of solar variation, greenhouse gases, and other factors to a complicated cycle of temperature changes is a "giant statistical problem," North says. "You've got to use pretty sophisticated statistical methods in these analyses, and the results can fool you really easily if you don't understand what you're looking at."

Still, he notes, there's overwhelming evidence that Earth is warming. His models show with a 75 percent certainty that solar variation is contributing 10 to 20 percent of the temperature rise of a fraction of a degree. This small change, he contends, is useful only as a theoretical test for climate models.

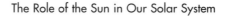

Other scientists also have a tough time believing that just a small variation in solar irradiance over an 11-year period could have a strong effect on climate. "They've said, 'It just can't be,'" says NOAA's Viereck. "But even some of the strongest nay sayers are starting to recognize that there's something there."

One man who may convert more naysayers to yea-sayers is John K. Lawrence, a physicist at California State University in Northridge. Lawrence and his colleagues used what he calls "an extremely crude model" of three interrelated equations to simulate the flow patterns in the atmosphere at middle latitudes.

Although simple, the model incorporates all of the basic physics equations needed to describe one aspect of the atmospheric flow around Earth. Researchers using this model can add input that simulates variations in solar radiation.

The model calculates the average speed of the westerly flow of winds in the atmosphere as a function of latitude. This number provides a crude estimation of the speed of the jet stream that drives weather systems from west to east.

"It's a good general circulation model," says Lawrence. "But if you want to predict the weather in Scotland 2 years from now, this model won't do it."

What the model does do, Lawrence says, is show that even a simple system of equations can answer many of the criticisms directed against those who say solar variation is an important factor in explaining significant global temperature changes.

For one thing, the model exhibits what physicists call chaotic behavior—even small changes in the inputs to complex systems can cause large variations in the answers (see "The Changes of Wind," [on the next page]). This, Lawrence notes, helps bolster the claim that a 0.1 percent variation in solar irradiance can, in fact, significantly affect climate.

Second, the model's calculations produce correlations that appear during the early phases of a simulation, disappear later in the simulation, and then reappear as anticorrelations. This matches the past behavior of the solar cycle. Between 1860 and 1920, cooler temperatures occurred when sunspot numbers were large. From the 1920s to the 1960s, there was no clear correlation between sunspot numbers and temperature. But after 1960, increased sunspots correlated with higher temperatures, Lawrence notes.

Finally, Lawrence's model allows the Northern and Southern Hemispheres to fluctuate independently of one another and to have different correlations to the solar variation, as has been observed.

"Most scientists try to explain these things by creating incredibly detailed models and then solve them by brute force with lots of computer power," says Lawrence. "We're saying it's plausible to get big, complex effects from a remarkably simple model."

For example, small variations in solar irradiance may drive big changes in climate. He adds: "Whether the climate actually works this way is another story."

The Changes of Wind

Output of a climate model that suggests even small variations in solar irradiance may drive big changes in climate: The [second image in the original article] shows a plot of average westerly wind speed as a function of the heating contrast between high and low latitudes (horizontal axis) and eddy forcing (vertical axis), a parameter that can represent the heating differences between land and sea. For some combinations of the two parameters, such as those represented in the upper left corner of the larger image, average wind speeds remain relatively stable. For other combinations, such as those shown in the lower left corner (a region shown magnified in the inset at left), even minuscule changes in input can cause significant fluctuations in average wind speed.

References

Evans, W. F. J. 2001. A comparison of the relative contributions of solar variability and C02. 81st Annual Meeting of the American Meteorological Society. Jan. 14–19. Albuquerque.

Lawrence, J. K., A. C. Cadavid, and A. Ruzmaikin. 2000. The response of atmospheric circulation to weak solar forcing. *Journal of Geophysical Research* 105 (Oct. 27):24839.

Marsh, N. D., and H. Svensmark. 2000. Low cloud properties influenced by cosmic rays. *Physical Review Letters* 85 (Dec. 4):5004–5007. Abstract available at http://link.aps.org/abstract/PRL/v85/p5004.

Perry, C. A., and K. J. Hsu. 2000. Geophysical, archaeological, and historical evidence support a solar-output model for climate change. *Proceedings of the National Academy of Sciences* 97 (Nov. 7):12433–12438. Abstract available at http://www. pnas. org/cgi/content/abstract/97/23/12433.

Powell, A. M., P. A. Zuzolo, and B. J. Zuzolo. 2001. Investigation of a proposed solar-terrestrial relationship with potential monthly and decadal implications. 81st Annual Meeting of the American Meteorological Society. January 14–19. Albuquerque.

Reddy, R. S. 2001. An evidence of a 11-year solar cycle in the Alaskan climate. 81st Annual Meeting of the American Meteorological Society. Jan. 14–19. Albuquerque.

Stott, P. A., *et al.* 2000. External control of 20th century temperature by natural and anthropogenic forcings. *Science* 290 (Dec. 15):2133–2137. Abstract available at http://www.sciencemag.org/cgi/content/abstract/290/5499/2133.

Udelhofen, P. M. 2001. Effect of the solar cycle cloud cover variations over the United States. 81st Annual Meeting of the American Meteorological Society. Jan. 14–19. Albuquerque.

Zwiers, F. W., and A. J. Weaver. 2000. The causes of 20th century warming. *Science* 290 (Dec. 15):2081–2083. Summary available at http://www.sciencemag. org/cgi/content/summary/290/5499/2081.

Our views and perspectives are shaped by our life experiences. Scientists face similar limitations. If human activities are influencing Earth's response to natural solar variation, it is very difficult to detect because we exist within an imperfect framework where pollution does exist and may or may not be affecting solar models or other study results.

To bypass this obstacle, researchers are analyzing what the climate was like thousands of years ago, long before the advent of cars and other fossil-fuel machines. Some researchers even study what the climate was like before the evolution of humans. In this case, Charles A. Perry and Kenneth J. Hsu look at models for Earth's climate 9,000 years ago. Patterns developed from these models suggest that we should be in a period of gradual cooling now, with a return to a mini ice age within the next 500

years. Since just the opposite seems to be happening, the findings could suggest that forces other than natural solar output may be influencing today's climate change. —JV

"Geophysical, Archaeological, and Historical Evidence Support a Solar-Output Model for Climate Change"
by Charles A. Perry and Kenneth J. Hsu
Proceedings of the National Academy of Sciences, **November 7, 2000**

The debate on the cause and the amount of global warming and its effect on global climates and economics continues. As world population continues its exponential growth, the potential for catastrophic effects from climate change increases. One previously neglected key to understanding global climate change may be found in examining events of world history and their connection to climate fluctuations.

Climate fluctuations have long been noted as being cyclical in nature, and many papers have been published on this topic (1). These fluctuations also can be quite abrupt (2) when climate displays a surprisingly fast transition from one state to another. Possible causes of the cyclic variations and abrupt transitions at different time intervals have been theorized. These theories include internal drivers such as CO_2 concentrations (3), ocean temperature and salinity properties (4), as well as volcanism and atmospheric-transmissivity variations (5). External drivers include astronomical factors such as the Milankovitch orbital parameters (6), which recently have been challenged (7), and variations in the Sun's energy output (8–10).

The most direct mechanism for climate change would be a decrease or increase in the total amount of radiant energy reaching the Earth. Because only the orbital eccentricity aspect of the Milankovitch theory can account for a change in the total global energy and this change is of the order of only a maximum of 0.1% (11), one must look to the Sun as a possible source of larger energy fluctuations. Earth-satellite measurements in the last two decades have revealed that the total energy reaching the Earth varies by at least 0.1% over the 10- to 11-year solar cycle (12). Evidence of larger and longer term variations in solar output can be deduced from geophysical data (13–17).

In an extensive search of the literature pertaining to geophysical and astronomical cycles ranging from seconds to millions of years, Perry (18) demonstrated that the reported cycles fell into a recognizable pattern when standardized according to fundamental harmonics. An analysis of the distribution of 256 reported cycles, when standardized by dividing the length of each cycle, in years, by 2^N (where N is a positive or negative integer) until the cycle length fell into a range of 7.5 to 15 years, showed a central tendency of 11.1 years. The average sunspot-cycle length for the period 1700 to 1969 is also 11.1 years (19). In fact, the distribution of the sunspot cycles is very nearly the same as the distribution of the fundamental cycles of other geophysical and astronomical cycles. Aperiodicity of the cycles was

evident in two side modes of 9.9 and 12.2 years for the geophysical and astronomical cycles and 10.0 and 12.1 years for the sunspot cycle. The coincidence of these two patterns suggests that solar-activity cycles and their fundamental harmonics may be the underlying cause of many climactic cycles that are preserved in the geophysical record. Gauthier (20) noted a similar unified structure in Quaternary climate data that also followed a fundamental harmonic progression (progressive doubling of cycle length) from the 11-year sunspot data to the major 90,000-year glacial cycle.

Solar-Output Model

A simple solar-luminosity model was developed to estimate total solar-output variations throughout an entire glacial cycle of approximately 90,000 years (21). The model summed the amplitude of solar variance for each harmonic cycle from 11 to 90,000 years with the cumulative amplitude of 0.08 % for the 11-year cycle and 0.62 % for the 90,000-year cycle (18). The summation of the amplitudes of each of 13 individual harmonic cycles resulted in an estimate of the relative change of solar output through a full glacial cycle (full glacial through the interglacial and back to full glacial). Analysis of the ratio of oxygen 16 to oxygen 18 ($O^{16/18}$) in deep sea cores shows that the oceans have warmed and cooled in concert with the growth of continental glaciers. Emiliani (22) determined $O^{16/18}$ ratios, representing the last 720,000 years, show eight distinct glacial cycles averaging approximately 90,000 years in length. When the $O^{16/18}$ ratios for each of the eight cycles are standardized to cover a period of

90,000 years, they significantly correlate with the modeled solar-output amplitudes. Correlation coefficients between $O^{16/18}$ and modeled solar output for these eight glacial cycles averaged $r = 0.64$ (18).

Examination of the unstandardized, individual cycles in $O^{16/18}$ ratios of the deep-sea cores revealed a tendency for the shorter cycles to represent warmer conditions and the longer cycles to represent colder conditions. A relation among solar magnetic intensity, total solar-energy output, and the length of the solar cycle was speculated by Perry (23) and was later supported by Friss-Christensen and Lassen (8) with their comparison of global air temperatures and solar-cycle length. Accordingly, it may be inferred that during periods of global warmth (interglacial) the basic fundamental solar cycle would expected to be shorter, possibly averaging 10 years, whereas during global chill (glacial), the basic cycle would expected to be longer, possibly closer to 12 years. These two values correspond to the two side modes of the distribution of the basic cycles in geophysical data mentioned previously.

The solar-output model was modified to allow the basic cycle to vary from a 12-year basic cycle during the last 30,000 years of the last glacial cycle up to the Pleistocene/Holocene boundary [approximately 9,000 years before present (BP)] to a 10-year basic cycle onward for another 19,000 years [to 10,000 years after present (AP)]. This 50,000-year period includes both the full glacial and altithermal interglacial periods in the approximate 90,000-year cycle. The two modeled time periods were spliced together at the point where the change in solar output was 0% from the 90,000-year average.

Figure 1 [in the original article] shows the resultant solar output model variations of luminosity compared with selected events deduced from the geophysical records and archaeological evidence during the late Pleistocene and Holocene.

Model Timeline Calibration

To initially calibrate the model, the rapid warm-up of the Pleistocene/Holocene time boundary was determined to be the most significant climactic event in the last 40,000 years. The date at the Pleistocene/Holocene boundary has been placed at approximately 10,000 years BP (6), although there is some debate on when the greatest warm-up occurred. The sea-level curve of Ters (24) shows that the oceans were rising very rapidly by 9,000 BP. It is at this point in time that the solar-output model is date calibrated (point A, Fig. 1 [in the original article]). In further comparison of the sea-level fluctuations during the Holocene, the shape of the Ters sea-level curve (24) generally resembles that of the solar-output model. Other dated events in the very late Pleistocene can be compared with the solar-output model. Larson and Stone (25) have dated the farthest full glacial advance at 21,800 BP ([point B, Fig. 1 in the original article]), whereas Shackleton and Hall (26) and Woillard and Mook (27) have independently dated the coldest temperatures of the last full glacial period as being 16,000 BP (point C, Fig. 1 [in the original article]). Both periods coincide with low points in the solar-output model. Later, the warm glacial periods were interrupted by brief and sudden cold periods of the Older Dryas centered near 12,100 BP (point D, Fig. 1 [in the original article]) and the Younger Dryas near 10,300 BP (28) (point D, Fig. 1 [in the original article]) which also coincide with low points in the solar-output model.

Comparison of Modeled Solar Output and Archaeological and Historical Evidence

A warming period from 33,000 to 26,000 years BP (bar at F, Fig. 1 [in the original article]) during the last continental glaciation may have allowed the Cro-Magnon Man to migrate northward and populate Europe by either acculturation (29) or eradication of the resident Neanderthals. Linguistic (9) and genetic evidence (30) suggest that the Americas were populated during the very late Pleistocene by several waves of migrations from Asia, one of which may have occurred approximately 20,000 years BP (point G1, Fig. 1 [in the original article]). These migrations possibly could have occurred as early as 32,000 years BP, but other groups may have migrated during warm-ups near 15,000 years BP (point G2, Fig. 1 [in the original article]) and near 11,000 years BP (point G3, Fig. 1 [in the original article]) just before the land bridge between Asia and Alaska was finally inundated by the rapidly rising sea levels.

The solar-output model can be evaluated more accurately with dated events in the Holocene. Fig. 2 [in the original article] is a more detailed depiction of the solar-output model representing the period 14,000 years BP to 2,000 years AP. A notable feature is the approximate 1,300-year little-ice-age cycle that is apparent throughout the Holocene in the Ters sea-level curve and in the output model. This also coincides with historical events

interpreted by Hsu (9) as being related to periodic climate changes, as evidenced by the records of history and natural science. The Older Dryas at 12,100 years BP and the early phase of the Younger Dryas at 10,800 years BP (28) correspond well with the little-ice-age cycle of approximately 1,300 years throughout the Holocene. However, there is one exception to the regular little-ice-age cycle. There is no evidence of a cold period near 9,500 years BP in the sea-level data. The solar-output model near 9,500 years BP also does not show a decrease in solar output but instead a sharp increase. During this period the positive amplitude of all of the other 12 harmonic cycles completely overwhelms the approximate 1,300-year little-ice-age cycle. After missing a little-ice-age period during the Pleistocene/Holocene transition it returned on time near 8,200 years BP in both the solar-output model and the geophysical climate record.

The "Global Chill" centered near 8,200 years BP on the solar-output model is reflected in a small dip in the otherwise steady rise in the Ters sea-level curve. Later, there is evidence for two "Sahara Aridity" cold periods, one centered near 7,000 years BP (point I, Fig. 2 [in the original article]) and another at 5,500 years BP (point J, Fig. 2 [in the original article]), which prompted a great migration of a pastoral civilization from the now Sahara Desert to the Nile River Valley in Egypt (9). The "4,000 BP Event" that in fact prevailed from 4,400 to 3,800 BP (point K, Fig. 2 [in the original article]) (9) may have been the coldest period since the Younger Dryas cold period. The "Centuries of Darkness" from 3,250 to 2,750 years BP included the downfall of the great empires of the Bronze Age (9). Another little ice age occurred during the period from 2,060 to 1,400 years BP [60 B.C. to A.D. 600] called the "Migration of Nations," when at its coldest point, the Germanic tribes overran the Roman Empire and the northern Asiatic tribes overran the Chinese Empire (9). Concurrently in Central America, the cooler climate may have allowed the Mayan civilization to prosper. The cooling of this area of the tropics forced the mosquitoes that carried malaria to move farther south, allowing extensive farming and the construction of cities. When world climate warmed again, the Maya had to abandon their fields and cities and were forced to migrate northward to escape the returning malaria scourge (9).

The most recent of the climactic cooling periods was experienced during 720 to 140 years BP (A.D. 1280–1860) when the climate worldwide was probably the coldest since the continental glaciers melted 10,000 years ago and is referred to as the "Little Ice Age" (31). Although the cold periods of the little ice ages vary in length and severity, they seem to track the solar-output model reasonably well. As a result of cold and drought and war, human population probably declined during each of these little ice ages.

In contrast, warmer climates were accompanied by more rain, longer growing seasons, more crops, and more land to settle on. Civilizations prospered and great human achievements were attained. A similar cycle length of approximately 1,300 years between warm periods extends through the Holocene. Rapid warming about 7,600 years BP (point O, Fig. 2 [in the original article]) coincides with the flooding of the Black Sea basin (32) as sea levels rose from the melting of the remnants of the continental glaciers, Antarctic glaciers, and the many alpine glaciers. Nearly 1,300 years later between 6,500 and 6,000 years BP (point P, Fig. 2 [in the original article]), Brittany megalithic monuments were built (9). The climate of the British Isles had returned to a favorable level, allowing prosperity to return after the little ice age near 6,900 years BP. The warmer climate and prosperity allowed people time to haul huge blocks of rock many miles to construct great monuments. This civilization eventually was lost during the following little ice age during which the Man-in-Ice was buried under the first alpine glacier of the Dolomites near 5,300 years BP (9). By 5,000 years BP the warmth returned, and the first Egyptian Empire was beginning to flourish, and near 4,500 years BP, the Great Pyramids were constructed (point Q, Fig. 2 [in the original article]) (9). In India, the Harappa civilization also flourished during these warmer and wetter times only to be eradicated by conditions brought on by extreme drought during the long-lasting cold "4,000 BP Event" (9, 33).

The 4,000 BP Event was perhaps the most influential little ice age in recorded history. Global cooling started as early as 4,400 BP and ended some 600 years later. Archaeological evidence (33) indicates that the years 4,200–3,900 BP were coldest and most arid in western Asia. The cooling signifies the change from the early Holocene climactic Optimum to late Holocene alternations of little climactic optimums and little ice ages (9). Varves from Swiss lakes indicate that the alpine glaciers became widespread during this first of the historical little ice ages. An important historical consequence of the 4,000 BP Event was the migration of the Indo-European peoples from northern Europe, to Greece, to southern Russia, to Anatolia, to Persia, to India, and to Xinjiang in northwest China (9).

The next warm period ushered in the Bronze Age, which began about 3,800 years BP (point R, Fig. 2 [in the original article]); this probably was the most favorable climactic period of the Holocene and is also referred to as the Holocene Maximum (9). People migrated northward into Scandinavia and reclaimed farmland with growing seasons that were at that time probably the longest in more than 2 millennia. The great Assyrian Empire, the Hittite Kingdom, the Shang Dynasty in China, and the Middle Egyptian Empire flourished (9). The Bronze Age came to an end with the "Centuries of Darkness" chill, but warming returned during the "Greco-Roman Age" from 2,750 to 2,060 years BP (9) (point S, Fig. 2 [in the original article]). During this period, philosophy made its first important advances with the thoughts of the great Aristotle. However, the climate cooled again between 2,000–1,400 years BP and the Roman Empire came to an end.

The next warm period was known as the Medieval Optimum (9) (point T, Fig. 2 [in the original article]), which was just beginning near 1,400 years BP and lasted until the Little Ice Age began about 700 years later. Currently, the Earth is enjoying the latest warm period (point U, Fig. 2 [in the original article]) which has been underway for almost two centuries.

The solar-output model and selected events are expanded in (Fig. 3 [in the original article]) for the period after about 2,300 years BP. The initial time calibration remains the same, only the resolution of the solar-output model and the selected events is increased. To avoid confusion of dates, the Gregorian calendar system will be used for the next section where 2,000 years BP coincides with 1 B.C.

At the time the Mayan civilization abandoned their great cities in Central America and migrated northward into the Yucatan Peninsula (A.D. 500), the Vikings were gaining power in northern Europe. By A.D. 1000 (bar T1, Fig. 3 [in the original article]), they had discovered Greenland and had traveled on to Newfoundland. Grain was grown in Greenland, and northern Europe prospered as the first millennium (A.D.) came to an end (34). However, after A.D. 1000 the climate in the northern latitudes began to slowly deteriorate with the cooler centuries of the Medieval Optimum during the 1000s and 1100s (bar T2, Fig. 3 [in the original article]). A brief return to a warm climate near A.D. 1200 (bar T3, Fig. 3 [in the original article]) coincided with an increase in the building of cathedrals in Europe.

After A.D. 1200, the climate began to cool more rapidly. The frozen harbors of Greenland failed to open in the summer, and trade with Europe dwindled rapidly (34). Thirty-five hundred miles to the southwest, the Anasazi civilization in the southwestern part of North America left the tranquility of large pueblos for the protection of the cliff dwellings. Maize became more difficult to grow as the continuing lack of precipitation allowed the desert to return to the American Southwest (9). By A.D. 1400 (bar N1, Fig. 3 [in the original article]), Europe's contact with Greenland had been lost, and the Anasazi had totally abandoned even the cliff dwellings. A slight warm-up about A.D. 1500 (bar N2, Fig. 3 [in the original article]) allowed the return of ships to Greenland, but they found that the stranded Viking population had starved to death. Examination of their cemeteries showed that with time the graves became shallower as the permafrost returned (34). The 1500s was only a brief respite as the coldest times of the Little Ice Age were yet to come in the 1600s (bar N3, Fig. 3 [in the original article]).

Half a world away in the tropical Pacific Ocean a similar saga unfolded. During the Greco-Roman climactic optimum, the Polynesians migrated across the Pacific from island to island, with the last outpost of Easter Island being settled around A.D. 400 (35). Between A.D. 1000 and 1350 the people moved to the ocean shores great blocks of stone carved into Moais. Moai building ceased around A.D. 1350 as famine spread across the island. Did the people of Easter Island eventually fall into cannibalism by the late 1600s (35) as a result of environmental degradation by overpopulation, or was it a major change in global climate caused by a decrease in solar output that converted their home from a wet tropical island into a desert island during the Little Ice Age?

Fluctuations of the Sun's intensity are recorded by the amount of carbon 14 in well-dated tree rings during this period (36). Carbon-14 is produced in the atmosphere by cosmic rays that are less abundant when the Sun is active and more abundant when the Sun is less

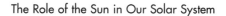

active. The fluctuations of carbon-14 in dated tree rings are inverted in Fig. 3 [in the original article] for easier comparison to the solar-output model. The carbon-14 trace corresponds well with the solar-output model and selected events since the Medieval Optimum (about A.D. 1100), which further supports the hypothesis that the Sun is varying its energy production in a manner that is consistent with the superposition of harmonic cycles of solar activity.

Great civilizations appear to have prospered when the solar-output model shows an increase in the Sun's output. Increased solar output would have caused the atmosphere and oceans to warm, therefore increasing the amount of water vapor in the atmosphere and causing increased precipitation. Growing seasons in the more polar latitudes became extended. Lands marginal for human habitation became favorable to support a growing population as deserts became wetter and the subArctic became warmer. Great civilizations appear to have declined when the modeled solar output declined. Severe and long-lasting droughts came to the steppes, winters in the subArctic became fierce, and growing seasons shortened. Similar processes currently are evident on a smaller scale. In the central parts of North America, droughts occur after the Sun's output has decreased slightly over a period of several of years, and when the Sun's output increases, an abundance of moisture follows in several years (10).

Current global warming commonly is attributed to increased CO_2 concentrations in the atmosphere (3). However, geophysical, archaeological, and historical evidence is consistent with warming and cooling periods during the Holocene as indicated by the solar-output model. The current warm period is thought to have not reached the level of warmth of the previous warm period (A.D. 800–1200), when the Vikings raised wheat and livestock in Greenland. Therefore, the magnitude of the modern temperature increase being caused solely by an increase in CO_2 concentrations appears questionable. The contribution of solar-output variations to climate change may be significant.

Extrapolation of Solar Output

Because of the favorable agreement between the solar-output model fix and past climactic events, it is possible to speculate on long-range projections of global climate. Projections of future solar-luminosity variations provided by the solar-output model are shown as a dashed line in Figs. 1–3 [in the original article]. Extrapolation of the model shows that the current warm period may be ending and that the Earth's climate may cool to conditions similar to the Little Ice Age by A.D. 2400–2900. The total decrease in solar output may be only about half the decrease experienced from A.D. 900 to 1700. The model shows a slight cooling during A.D. 2000–2100, possibly leading to more severe droughts in the African Sahel than were seen in the minor global cooling that occurred in the 1970s. The following three centuries (A.D. 2100–2400) show a gradual temperature decrease toward another little ice age. However, by the middle of the third millennium A.D. (1,500 years from now), the model shows a return to the altithermal conditions similar to the Holocene Maximum that occurred during the

Bronze Age that began about 3,800 years ago. Warmer and wetter conditions may prevail on a global scale, with the warmth possibly sufficient to melt the polar icecaps and flood the world's coastlines. There is evidence that this happened during the last interglacial period, when sea levels were 3–5 m above present sea levels (37).

The solar-output model timeline also can provide an estimate of the length of the current interglacial period. Marine cores have suggested that the average length of the interglacial periods is approximately 10,000 years (38). However, there is compelling continental evidence that the average interglacial period may be twice as long. A 500,000-year-long delta oxygen-18 time series obtained from vein calcite in Devel's Hole, Nevada, shows an average of 22,000 years for the length of the interglacial periods (39). Other continental and marine evidence (40) during the previous interglacial period ($\approx$130,000 to $\approx$110,000 years BP) show fluctuations toward glacial conditions that are consistent with the solar-output model. In the model, solar output is greater than 0% of the glacial-cycle average for 15,000 years, from 9,000 BP to 6,000 years from now (Fig. 1 [in the original article]). About 10,000 years from now, the model shows an end to the interglacial period and a gradual temperature decrease into glacial conditions punctuated by increasingly briefer warm periods and longer and more intense little ice ages.

Summary

The development and calibration of a solar-output model for climate are supported by geophysical, archaeological, and historical evidence from the last full glacial

Pleistocene (30,000 years BP) through the current Holocene interglacial to the present. The solar-output model is based on a superposition of a fundamental harmonic progression of cycles beginning at 10 and 12 years and progressing to the 13th harmonic (90,000-year cycle), which is approximately equal to the average continental glacial cycle. This model was date calibrated to the Pleistocene/Holocene boundary at 9,000 years BP and compared with geophysical records of sea level, carbon-14 production, oxygen 16/18 ratios, and other geologic evidence of climate fluctuations. The approximate 1,300-year little-ice-age cycle and intervening warmer periods agree with archaeological and historical evidence of these cold and warm periods. Throughout history, global warming has brought prosperity whereas global cooling has brought adversity.

The solar-output model allows speculation on global climactic variations in the next 10,000 years. Extrapolation of the solar-output model shows a return to little-ice-age conditions by A.D. 2400–2900 followed by a rapid return to altithermal conditions during the middle of the third millennium A.D. This altithermal period may be similar to the Holocene Maximum that began nearly 3,800 years ago. The solar output model suggests that, approximately 20,000 years after it began, the current interglacial period may come to an end and another glacial period may begin.

1. Rampino, M. R., Sanders, J. E., Newman, W. S. & Konigsson, L. K. (1987) *Climate History, Periodicity, and Predictability* (Van Nostrand Reinhold, New York).
2. Berger, W. H. & Labeyrie, L. D. (1987) *Abrupt Climate Change* (D. Reidel, Boston).

3. Manabe, S. & Wetherald, R. T. (1980) *J. Atmos. Sci.* 37, 99–118.
4. Broecker, W. S. (1995) *Sci. Am.* 273, 62–68.
5. Hartmann, D. L., Mouginis-Mark, P., Bluth, G. J., Coakley, J. A., Jr., Crisp, J., Dickinson, R. E., Francis, P. W., Hansen, J. E., Hobbs, P. E., Isacks, B. L., *et al.* (1999) in *EOS Science Plan: The State of Science in the EOS Program*, eds., King, M. D., Greenstone, R. & Bandeen, W. (Report No. NP-1998-12-069-GSFC), (National Aeronautics and Space Administration, Greenbelt, MD), pp. 339–378.
6. Imbrie, J. & Imbrie, K. P. (1979) *Ice Ages: Solving the Mystery* (Harvard Univ. Press, Cambridge).
7. Karner, D. B. & Muller, R. A. (2000) *Science* 288, 2143–2144.
8. Friss-Christensen, E. & Lassen, K. (1991) *Science* 254, 698–700.
9. Hsu, K. J. (2000) *Climate and Peoples: A Theory of History* (Orell Fussli, Zurich).
10. Perry, C. A. (2000) in *Proceedings of the 16th Annual Pacific Climate (PACLIM) Workshop*, May 24–27, 1999, Santa Catalina, CA, eds., West, G. J. & Buffaloe, L. (Tech. Report 65 of the Interagency Ecological Program for the Sacramento-San Joaquin Delta), (State of California, Dept. of Water Resources, Sacramento), pp. 161–170.
11. Milankovitch, M. M. (1941) *Canon of Insolation and the Ice-Age Problem* (Koniglich Serbische Akademie, Beograd).
12. Willson, R. C. & Hudson, H. S. (1988) *Nature (London)* 332, 810–812.,
13. Youji, D., Baorong, L. & Yongming, F. (1979) in *Weather and Climate Responses to Solar Variations*, ed. McCormac, B. M. (Colorado Associated Univ. Press, Boulder), pp. 545–557.
14. Reid, G. C. (1991) J. Geophys. Res. 96, 2835–2844.
15. Beer, J., Joos, F., Lukasczyk, Ch., Mende, W., Rodriguez, J., Siegenthaler, U. & Stellmacher, R. (1993) in *The Solar Engine and Its Influence on Terrestrial Atmosphere and Climate*, ed. Nesme-Ribes, E. (Springer, Berlin), pp. 221–234.
16. Vos, H., Sanchez, A., Zolitschka, B., Brauer, A. & Negendank, J. F. W. (1977) *Surv. Geophys.* 18, 163–182.
17. Hoyt, D. V. & Schattan, K. H. (1997) *The Role of the Sun in Climate Change* (Oxford Univ. Press, Oxford).
18. Perry, C. A. (1989) *A Solar Chronometer for Climate, Astronomical and Geophysical Evidence* (University Microfilms International, Ann Arbor, MI).
19. Cole, T. W. (1973) *Solar Phys.* 30, 103–110.
20. Gauthier, J. H. (1999) *Geophys. Res. Lett.* 26, 763–766.
21. Perry, C. A. (1994) in *Proceedings of the Oklahoma/Kansas TER-QUA Symposium* (Stillwater, OK), ed. Dort, W., Jr. (Transcript Press, Norman, OK), pp. 25–37.
22. Emiliani, C. (1978) *Earth Planetary Sci. Lett.* 37, 349–352.
23. Perry, C. A. (1990) in *Proceedings of the Conference on the Climate Impact of Solar Variability* (NASA Conference Publication 3086, Goddard Space Flight

Center, Greenbelt, MD, April 24–27, 1990), eds. Schatten, K. H., & Arking, A. (National Aeronautics and Space Administration, Langley, VA), pp. 357–364.

24. Ters, M. (1987) in *Climate History, Periodicity, and Predictability*, eds. Rampino, M. R., Sanders, J. E., Newman, W. S. & Konigsson, L. K., (Van Nostrand Reinhold, New York), pp. 204–336.

25. Larson, B. J. & Stone, B. D. (1982) *Late Wisconsin Glaciation of New England* (KendalyHunt, Dubuque, IA).

26. Shackleton, N. J. & Hall, M. A. (1983) in *Initial Reports of the Deep Sea Drilling Project*, eds. Cann. J. R. & Langseth, M. G. (Texas A&M University, College Station), pp. 431–441.

27. Woillard, G. M. & Mook, W. G. (1982) *Science* 215, 159–161.

28. Schove, D. J. (1987) in *Climate History, Periodicity, and Predictability*, eds. Rampino, M. R., Sanders, J. E., Newman, W. S. & Konigsson, L. K., (Van Nostrand Reinhold, New York), pp. 355–377.

29. d'Errico, F., Zilhao, J., Julien, M., Baffier, D. & Pelegrin, J. (1988) *Curr. Anthropol.* 39, Suppl., S1–S44.

30. Mendoza, D. H. & Braginski, R. (1999) *Science* 283, 1441–1442.

31. Schotteer, U. (1990) *Climate: Our Future* (Kummerly & Frey, Bern).

32. Ryan, W. B. F., Pitman, W. C., III, Major, C. O., Shimkus, K., Moskalenko, V., Jones, G. A., Dimitrov, P., Gorur, N., Sakinc, M. & Yuce, H. (1997) *Marine Geol.* 138, 119–126.

33. Weiss, H., Coury, M. A., Wetterstrom, W., Guichard, F., Senior, F., Meadow, R. & Curnow, A. (1993) *Science* 261, 995–1004.

34. Jones, G. (1968) *A History of the Vikings* (Oxford Univ. Press, Oxford).

35. McCall, G. (1995) *Pacific Islands Yearbook* (Fiji Times, Suva, Fiji), 17th Ed.

36. Damon, P. E. (1977) in *The Solar Output and its Variations,* ed. White, O. R. (Colorado Associated Univ. Press, Boulder), pp. 429–448.

37. Zhu, Z. R., Wyrwoll, K. H., Collins, L. B., Chen, J. H., Wasserburg, G. J. & Eisenhauer, A. (1993) *Earth Planetary Sci. Let.* 118, 281–293.

38. Slowey, N. C., Henderson, G. M. & Curry, W. B. (1996) *Nature (London)* 383, 242–244.

39. Winograd, I. J., Landwehr, J. M., Ludwig, K. R., Coplen, R. B. & Riggs, A. C. (1997) *Quat. Res.* 48, 141–154.

40. Kukla, G. J. (2000) *Science* 287, 987–988.

Reprinted with permission from *Proceedings of the National Academy of Sciences*, Vol. 97, No. 23, November 7, 2000, pp. 12433–12438. © 2000, National Academy of Sciences, U.S.A.

5 The Sun's Ultraviolet-B Radiation

The Sun releases waves of electromagnetic energy. These waves exist in several forms. The most obvious is visible light. Beyond what we can see and directly feel are other invisible waves. One of the most common is called ultraviolet-b (UVB) radiation. The term refers to the fact that the waves exist beyond the violet end of the light spectrum. It is sort of like an invisible extension of a rainbow.

Unlike rainbows, UVB waves often are unwelcome visitors to Earth. Like other damaging forms of radiation, UVB waves can damage DNA, which is the genetic material that guides most biochemical reactions in our bodies and those of other living creatures. DNA is like hardware in a computer. If hardware is damaged, the entire system can break down. Similarly, if DNA is damaged, it can lead to diseases, such as cancer, and even death. In the paper "Solar UVB-Induced DNA Damage and Photoenzymatic DNA Repair in Antarctic Zooplankton," researchers discuss how they have detected DNA damage in tiny *zooplankton located in the Antarctic. While these tiny creatures sometimes can repair the genetic damage, similar to how our immune systems can flush out viral agents, the overall zooplankton population and the species that feed upon it can suffer lasting problems related to UVB radiation exposure.* —JV

"Solar UVB-Induced DNA Damage and Photoenzymatic DNA Repair in Antarctic Zooplankton"
by Kirk D. Malloy, Molly A. Holman, David Mitchell, and H. William Detrich III
Proceedings of the National Academy of Sciences, February 1997

Abbreviations: CPD, cyclobutane pyrimidine dimer; PER, photoenzymatic repair; NER, nucleotide excision repair.

The concentration of stratospheric ozone has decreased significantly during the past two decades, the result of catalytic destruction of ozone mediated by the photodegradation products of anthropogenic chlorofluorocarbons (1–2). Ozone depletion has been most dramatic over Antarctica, where ozone levels typically decline > 50 % during the austral spring "ozone hole" (3–5). Elsewhere, ozone concentrations have fallen gradually at temperate latitudes (6, 7), and further depletion over a broader geographical range is anticipated during the next 25–100 years (7, 8). Atmospheric ozone strongly and selectively absorbs solar UVB (290–320 nm), thus reducing the intensity of the most biologically damaging solar wavelengths that penetrate the atmosphere (9), and decreased stratospheric ozone has been linked directly to increased UVB flux at the earth's surface (10). UVB penetrates to ecologically significant depths (20–30 m) in the ocean at intensities that cause measurable biological damage (11–16). Therefore, the fitness of marine organisms in coastal regions and the upper photic zone of open oceans may be affected deleteriously by the projected long-term increase in UVB flux (16–20).

Although UVB damages most biological macromolecules, including lipids, proteins, and nucleic acids (21, 22), the principal cause of UV-induced mutation, degeneration, and/or death in animal cells is modification of DNA (23). The most significant DNA lesions generated by UVB are cyclobutane pyrimidine dimers (CPDs), which constitute ≈70–90% of the aberrant DNA photoproducts (23). CPDs increase linearly with UVB exposure (23, 24), but the dose-response relationship varies significantly between different species and taxa (25–27). Nevertheless, the CPD burden resulting from sublethal doses of UVB may inhibit embryonic and larval development and ultimately decrease survival, by slowing transcription and mitosis and by imposing energetic costs associated with DNA repair (28).

The impact of elevated UVB on marine ecosystems has been documented most extensively for the primary producers of polar latitudes (refs. 13 and 29; reviewed in ref. 30). Primary productivity in the Southern Ocean declines by as much as 15% in areas affected by the ozone hole (13). Survival of and photosynthesis by Antarctic diatoms are reduced by UVB, although the susceptibility of these single-celled algae to UVB-induced DNA damage varies significantly, depending upon cell surface:volume ratios, pigmentation, and DNA repair rates (25). UVB has also been shown to inhibit nitrogen uptake by phytoplankton from the North Atlantic (31). Although previous investigators have strongly suggested that elevated UVB flux may perturb marine ecosystems as a whole (32, 33), its effects on marine organisms other than primary producers have not yet been systematically investigated. Bothwell *et al.* (34), after demonstrating that herbivorous grazers were the keystone species affected by UVB in shallow temperate streams, concluded that "predictions of the response of entire ecosystems to elevated UVB cannot be made on single trophic-level assessments."

Most organisms possess defense mechanisms that act either to prevent UVB-induced DNA damage (via behavioral, physical, and/or chemical strategies) or to repair it after it has occurred. The two primary repair systems are photoenzymatic repair (PER; "light repair") and nucleotide excision repair (NER; "dark repair") (25). PER repairs primarily CPDs via the enzyme photolyase, which uses near-UV light (320–400 nm) as its source of energy. NER, by contrast, involves multiple proteins (e.g., UVR-A, -B, and -C in *Escherichia coli*), does not require light for catalysis and repairs preferentially the 6–4 pyrimidine–pyrimidinone photoproduct and the Dewar pyrimidinone (35).

To assess the potential impact of ozone depletion and increased UVB intensities on populations of marine heterotrophs in Antarctica, we analyzed field-collected zooplankton for UVB-induced DNA damage (CPDs) during the ozone hole (October–November) of 1994. Because DNA damage and repair occur contemporaneously in the field, we also evaluated the ability of three Antarctic heterotrophs [two teleosts, *Notothenia coriiceps* (Nototheniidae; rockcods) and *Chaenocephalus aceratus* (Channichthyidae; icefishes) and one species of krill, *Euphausia superba*] to repair UV-induced DNA damage at ambient temperatures (-1 to $+1°C$). Finally, to evaluate the relative efficiency of DNA repair in these cold-adapted Antarctic species, we measured DNA repair rates of a temperate, eurythermal teleost (*Fundulus heteroclitus*) at three acclimation temperatures. Our results indicate that Antarctic zooplankters accumulate significant CPD levels during periods of increased UVB flux, that repair of this damage is mediated largely by the photoenzymatic repair system, and that DNA repair capacity is highest for species whose eggs and larvae occupy the water column during the austral spring. Although the reduction in zooplankton fitness resulting from this DNA damage is unknown, we suggest that increased solar UVB flux may reduce recruitment and trophic transfer of productivity by affecting both the heterotrophic species and the primary producers of the Antarctic marine ecosystem.

Materials and Methods

Zooplankton Collections. Macrozooplankton was collected from Antarctic surface waters (0–35 m deep) of the Palmer Archipelago (see Fig. 1 [in the original article] for sampling locations) during the austral spring of 1994

(*R/V Polar Duke* cruise 94-10, October 13–November 16). Specimens included fish larvae, fish eggs (unhatched, late-somitic developmental stages), chaetognaths, and annelids. Each sampling event was conducted for 1–2 h, during daylight hours (0600–1800), with a 2 × 3 m rectangular midwater trawl equipped with a 3-mm nylon delta mesh net. Each sampling event fished at a constant depth, estimated by wire angle and length of wire deployed. Organisms from net samples were identified (36), sorted by taxa, frozen in liquid nitrogen, and stored in the dark at −70°C until analysis for CPDs.

Ethanol-preserved *Notothenia larseni* larvae were obtained from Richard Radtke (University of Hawaii) to represent similar macrozooplankton collected at similar depths (0–45m) during "normal" ozone conditions (December 1988–January 1989). These samples, derived from the Meteor Cruise ME11/4 (37), were stored in UV-absorbing glass vials before use.

UVB Measurements. Cumulative daily solar UVB (290–320 nm) flux, weighted to a biological action spectrum determined for larval fish (38), was measured by the National Science Foundation Monitoring Station at nearby Palmer Station (data provided by Biospherical Instruments, San Diego). Hunter's weighted UVB flux ($\mu W/cm^2$), summed for each day (0000–2359 h) of cruise 94-10, was used as an estimate of biologically damaging surface UVB flux.

Quantitation of Photoproducts. Genomic DNA was purified from individual fish larvae, chaetognaths, or annelids and from samples (n = 78) of pooled icefish eggs (2–4 eggs per sample gave 15–50 µg of DNA per sample), by the method outlined in ref. 39. Ethanol-preserved fish larvae were rehydrated in TE buffer 10 mM•Tris HCl/1 mM EDTA, pH 8 for 24 h at 4°C before extraction of DNA. Duplicate aliquots of DNA (5 µg) from each sample were analyzed for CPD content by the competitive radioimmunoassay of Mitchell *et al.* (40). Briefly, unlabeled, heat-denatured specimen DNA, ^{32}P-labeled, UVB-irradiated poly(dA): poly(dT) competitive antigen (10 pg; 5×10^8 cpm/µg by random priming; 800 kJ/m^2 at 320 nm), and carrier salmon sperm DNA (5 µg/ml) were mixed, and rabbit antiserum (1:1000 dilution) specific for UV-damaged DNA was incubated with the sample for 1–2 h at 37°C. Immune complexes were collected by precipitation overnight at 4°C with goat anti-rabbit IgG and carrier IgG followed by centrifugation. After washing, pellets were dissolved in NCS II solubilizer (Amersham), and radioactivity was quantified by liquid scintillation counting in Scintiverse (Fisher). Specimen DNA damage (CPDs per Mb) was evaluated by use of a standard curve generated with unlabeled DNA containing known quantities of CPDs.

DNA Repair Measurements. To determine the relative importance of PER and NER repair systems in Antarctic marine heterotrophs, we measured light and dark DNA repair rates of juvenile specimens of two abundant fish species (the rockcod *N. coriiceps* and the icefish *C. aceratus*) and of adult specimens of krill (*E. superba*). The fishes, whose embryonic and larval stages differ in seasonal exposure to solar illumination (see *Results*), were collected during *R/V Polar Duke* cruises 94-10, 95-2, and 95-3 (October 1994–November 1994, March 1995, and May 1995, respectively). Krill were collected with a plankton net from surface waters near Palmer Station in February 1995. As a benchmark for comparison of DNA repair efficiency in these cold-living poikilotherms, we also evaluated repair rates of the eurythermal killifish, *F. heteroclitus*, at three acclimation temperatures (6, 10, and 25°C). This species [collected by minnow trap at Woods Hole, MA (June, 1995)] is an abundant component of most estuarine habitats on the East coast of the United States (41), with a latitudinal distribution from Newfoundland to south Florida (42). Its primary habitats are shallow saltmarsh creeks, in which it may encounter seasonal temperature ranges of −2°C in winter to +35°C in summer. *Fundulus* eggs typically develop out of the water for periods up to 4 weeks during the spring and summer, which makes them particularly vulnerable to UVB.

These organisms were acclimated to experimental temperatures for a minimum of 14 days and held in the dark for 24 h before exposure to 500-2500 J/m² UVB [generated by 5 × 8 W UVB lamps (Spectronic, Westbury NY), peak emission 312 nm, in a Fisher Biotech UV Crosslinker]. Immediately following irradiation, specimens used in the PER studies were exposed to continuous photoreactivating light (one General Electric 15-W Cool White fluorescent bulb at a distance of 25 cm) and then sacrificed at intervals for evaluation of CPD burden. DNAs from fin tissues of the Antarctic fishes, from fins and muscle-free skin of the killifish, and from whole krill were purified and analyzed for CPDs by radioimmunoassay as described above. PER rates were calculated by fitting the time course of CPD disappearance to the function $CPD = e-R^{(t+1)}$, where CPD is the percentage change in dimer concentration, R is the relative PER rate, and t is the post-UV irradiation time (in h). R was calculated by determining the least-squares slope of the logarithm-transformed relationship between CPD and time. Dark repair (NER) was evaluated with an identical protocol, but specimens were held in the dark for the duration of post-irradiation sampling.

Results

DNA Damage in Natural Populations of Antarctic Zooplankton. Eggs (unhatched, late-somitic stages) and larvae of the hemoglobinless icefish *C. aceratus* (Channichthyidae) predominated in the zooplankton collections, but chaetognaths and transparent planktonic polychaetes were also represented (Table 1 [in the original article]). During the period of sample collection, cumulative daily spectral irradiance values (290–320 nm) at nearby Palmer Station (Fig. 1 [in the original article]) peaked at 1308 µW/cm² (Hunter's UVB = 7.42) on October 31 but varied considerably [Hunter's UVB = 4.08 ± 1.85 (mean ± SD)]. Table 1 [in the original article] shows that a large proportion (> 50 %) of specimens in each taxon contained DNA damage. Ichthyofauna (eggs and larvae of *C. aceratus*) contained the highest levels of damage, both in CPD burden (average of 31–35 CPDs per Mb) and in the proportion of damaged specimens (> 67 %). In contrast, the *N. larseni* larvae, which are similar in size and coloration to the *C. aceratus* specimens but were collected during a period of lower UVB flux [daily fluxes in the range 413–666 µW/cm^2 (Hunter's UVB = 2.12–3.27), December 12–28, 1988] (T. Mestechkina, personal communication), had no detectable DNA damage. The sample sizes for chaetognaths and polychaetes were small, but half or more of these specimens also contained DNA damage, albeit at lower levels than those of the ichthyofauna. To our knowledge, this is the first demonstration that ozone depletion may be causing measurable DNA damage in natural populations of Antarctic heterotrophs.

DNA damage of *C. aceratus* eggs (daily means) correlated significantly with cumulative daily UVB flux ($P < 0.05$, Pearson's correlation) (Fig. 2 [in the original article]). Furthermore, mean CPD burden in *C. aceratus* eggs paralleled closely the biologically weighted surface UVB irradiance levels (Fig. 3A [in the original article]), which suggests strongly that DNA damage in these epipelagic zooplankton is largely attributable to solar UVB. CPD levels in *C. aceratus* larvae (heads and bodies analyzed separately), by contrast, tracked less closely with UVB intensity (Figs. 3B and C [in the original article]), perhaps due to differences in UV absorptivity and/or DNA repair capacity relative to eggs. No significant correlation was found between CPD content and depth of collection for any of the taxa examined (Pearson's $\alpha = 0.05$).

DNA Repair Rates of Antarctic Organisms. In both species of Antarctic fish and in krill, CPDs were repaired rapidly by PER at physiological temperatures (Table 2, Fig. 4A and B [in the original article]). For the eurythermal killifish, the rate of PER increased as a linear function of temperature (Fig. 4B [in the original article]), almost spanning from 6 to 25°C the range of rates observed for the Antarctic organisms at −1 to +1°C. Clearance of CPDs by NER, by contrast, was considerably slower (typically, < 15% reduction in CPDs after 24 h) for both the cold-adapted and temperate organisms (Table 2 [in the original article]). Thus, PER appears to be the major DNA repair system available to these marine species, but the ontogeny of PER and NER remains to be examined.

Within the Antarctic taxa, rates of DNA repair appear to correlate with the life histories of their embryonic stages. Particularly noteworthy are the high DNA repair capacities of icefish and krill, whose eggs and larvae remain in the water column throughout the austral spring and summer (44); at +1°C their PER rates are comparable to those observed for the killifish at 25°C (Fig. 4B [in the original article]). By contrast, PER rates for the rockcod, whose eggs and larvae develop in Antarctic surface waters during seasons of low solar illumination (austral fall and winter; ref. 44), are ≈ 50% smaller, matching the value extrapolated for the temperate fish (Fig. 4B, dotted line, [in the original article]). These differences are exemplified by the time required for 80% repair of initial CPD loads, which ranged from 4 to 5 h for the icefish and krill (+1°C) to > 15 h for the rockcod ($-1\,^{\circ}$C; Table 2 [in the original article]) If PER activities in icefish eggs and larvae are comparable to that measured in juveniles, then the daily tracking of DNA damage with UVB flux (Fig. 3 [in the original article]) is readily interpreted as the difference between temporally integrated damage and first-order CPD repair.

Discussion

The detrimental effects of increased solar UVB, caused by stratospheric ozone depletion, on single-celled phytoplankton have been known for several years (13, 29, 30). We show here that natural levels of solar UVB during ozone depletion may also cause measurable damage to multicellular organisms occupying higher trophic levels of the Antarctic marine ecosystem. The level of UVB-induced DNA damage measured in ichthyoplankton is greater than the lethal limit previously determined for Antarctic diatoms (> 15 CPDs per Mb) and is comparable to the lethal limit of DNA damage for cultured goldfish cells (20–100 CPDs per Mb) (25, 45). Despite substantial temporal changes in solar intensity, cloud cover, water column turbidity, and vertical mixing of the zooplankters during the sampling period, more than half of all specimens contained measurable DNA damage at levels that are probably physiologically relevant.

DNA damage was greater in the fish specimens than in the other taxa examined. In particular, the CPD content of icefish eggs closely followed daily UVB flux, which suggests that damage accrued during each 24-h period was repaired in less than 1 day (validating laboratory-derived DNA repair rates; see below). Icefish eggs, which are abundant, buoyant, and consistent in size, shape, and transparency, may therefore serve as useful biological indicators of the DNA-damaging effects of UVB in those zooplankters confined to Antarctic surface waters.

Icefish larvae, by contrast, showed patterns of DNA damage that correlated less well with daily UVB flux. Factors that may explain the distinct damage patterns observed for eggs and larvae include differences in: (i) UV absorptivity due to surface:volume ratio, pigmentation, content of UV-absorbing compounds, and/or mobility in the water column and (ii) DNA repair capacity. At present, the data required to discriminate between these possibilities is only suggestive. Icefish larvae contain moderate to high levels of UV-absorbing compounds (46), but the status of such materials in icefish eggs is unknown. DNA repair occurs at greater rates in young, undifferentiated cells (47), which suggests that DNA damage tracks ambient UVB more closely in icefish eggs than in larvae because eggs may be able to clear CPDs more rapidly. However, the expression and function of DNA repair systems during icefish development remains to be determined.

Assessment of the potential effect of elevated solar UVB on marine organisms requires consideration of light attenuation, mixing depth and sea state. Light intensity decreases exponentially with depth of the water column, yet biologically damaging levels of UV can penetrate to depths > 30m (11–16). Variability in sea state, turbidity, and vertical mixing/migration rates can influence substantially the actual damage accruing to marine organisms (11, 13, 14). For example, Jeffrey *et al.* (16) showed recently that sea state and vertical mixing rate are inversely proportional to the level of solar UV-induced DNA damage measured in marine bacteria in the Gulf of Mexico. Because we cannot reconstruct the precapture depth history of individual zooplankters in our field collections (depth of capture need not correspond to temporally weighted average depth of UVB exposure), it is not surprising that we found no relationship between collection depth and CPD load. Furthermore, diel and ontogenetic patterns of depth distribution are required to fully evaluate the impact of solar UVB on any marine taxon. We suggest, therefore, that buoyant, relatively nonmobile fish eggs may serve as effective passive *in situ* dosimeters, but accurate assessment of the effect of solar UVB on highly mobile taxa will require knowledge of vertical mixing rates and diel migratory behavior.

Antarctic fish and krill appear to rely primarily on PER, and therefore the single-enzyme system photolyase, to remove UVB-induced CPDs, as do several other taxa (25, 26, 40, 43). Each of the Antarctic species examined in this study demonstrated high repair capacities at the low ambient temperatures that characterize the Southern Ocean (-2 to $+2\,°C$), but rate differences possibly associated with the seasonality of embryogenesis were detected. Thus, the PER rates of the icefish *C. aceratus* and the krill *E. superba*, whose eggs and larvae are present in surface waters when solar illumination is high and ozone is depleted (austral spring and summer; ref. 48), are approximately twice as large as that of the rockcod *N. coriiceps*, whose eggs and larvae, spawning and developing during austral fall and winter, are exposed to much lower solar intensities. Although the data are preliminary, we hypothesize that Antarctic zooplankton whose early life history stages develop in the water column during periods of sustained, intense solar illumination, and hence are exposed to higher intensities of solar UV even in the absence of ozone depletion, may have increased capacities for DNA repair. Further evaluation of the PER systems of other Antarctic taxa would be required to test this proposal.

Comprehensive sampling of Antarctic zooplankton and field manipulation of their exposure to UVB, during both ozone-depleted and normal conditions, and determination of the effect of CPDs on the fitness of these organisms are required before the large-scale impact of ozone depletion on Antarctic marine heterotrophs can be fully evaluated. Our data suggest, however, that the effect on organisms occupying trophic levels other than primary producers may be substantial. Larval and adult fish, krill, copepods, and gelatinous zooplankton are all important components of the truncated trophic structure of the Southern Ocean (44, 48–51). They generally have transparent eggs, larvae, and/or adult stages that are pelagic, planktonic, and often remain in surface waters for 6–12 months (44, 49–54). Thus, these species are potentially vulnerable to DNA damage from the elevated UVB fluxes that now occur during the austral spring. The levels of DNA damage that we measured in natural populations of Antarctic zooplankton suggest that tissue damage accruing from UVB exposure may be of sufficient magnitude to decrease the fitness of developing eggs and larvae, thereby reducing recruitment. Thus, elevated solar UVB flux during the austral spring may have a substantial impact on populations of both primary producers and heterotrophs of the Antarctic marine ecosystem.

1. Anderson, J. G., Toohey, D. W. & Brune, W. H. (1991) *Science* 251, 39–46.
2. Schoeberl, M. R. & Hartmann, D. L. (1991) *Science* 251, 46–52.
3. Frederick, J. E. & Snell, H. E. (1988) *Science* 241, 438–440.
4. Brasseur, G. (1987) *Environment* 29, 6–11.
5. Solomon, S. (1990) *Nature (London)* 347, 347–354.
6. Madronich, S. & De Gruijl, F. R. (1994) *Photochem. Photobiol.* 59, 541–546.
7. Jones, A. E. & Shanklin, J. D. (1995) *Nature (London)* 376, 409–411.
8. Crawford, M. (1987) *Science* 237, 1557.
9. Molina, L. T. & Molina, M. J. (1986) *J. Geophys. Res. Atmos.* 91, 14501–14508.
10. Lubin, D., Frederick, J. E., Booth, C. R., Lucas, T. & Neuschuler, D. (1989) *Geophys. Res. Lett.* 16, 783–785.
11. Smith, R. C. & Baker, K. S. (1979) *Photochem. Photobiol.* 29, 311–323.
12. Calkins, J. & Thordardottir, T. (1980) *Nature (London)* 283, 563–566.
13. Smith, R. C., Prezelin, B. B., Baker, K. S., Bidigare, R. R., Boucher, N. P., Coley, T., Karentz, D., MacIntyre, S., Matlick, H. A., Menzies, D., Ondrusek, M., Wan, Z. & Waters, K. J. (1992) *Science* 255, 952–959.
14. Karentz, D., Cleaver, J. E. & Mitchell, D. L. (1991) *Nature (London)* 350, 28.
15. Calkins, J. (1982) in *The Role of Ultraviolet Radiation in Marine Ecosystems*, ed. Calkins, J. (Plenum, New York), pp. 169–179.
16. Jeffrey, W. H., Aas, P., Maille Lyons, M., Coffin, R. B., Pledger, R. J. & Mitchell, D. L. (1996) *Photochem. Photobiol.* 64, 419–427.
17. Worrest, R. C., Brooker, D. L. & Van Dyke, H. (1980) *Limnol. Oceanogr.* 25, 360–364.
18. Damkaer, D. M, Dey, D. B. & Heron, G. A. (1981) *Oecologia* 48, 178–182.
19. Worrest, R. C. (1982) in *The Role of Solar Ultraviolet Radiation in Marine Ecosystems*, ed. Calkins, J. (Plenum, New York), pp. 429–457.
20. Cullen, J. J. & Lesser, M. P. (1991) *Mar. Biol.* (Berlin) 111, 183–190.
21. Sauerbier, W. (1976) *Adv. Radiat. Biol.* 6, 49–106.
22. Kantor, G. J. & Hull, D. R. (1979) *Biophys. J.* 27, 359–370.
23. Carrier, W. L., Snyder, R. D. & Regan, J. D. (1982) in *The Science of Photomedicine*, eds. Regan, J. D. & Parrish, J. A. (Plenum, New York), pp. 91–112.
24. Setlow, R. B. (1974) *Proc. Natl. Acad. Sci. USA* 71, 3363–3366.

25. Mitchell, D. L. & Karentz, D. (1993) in *Environmental UV Photobiology*, eds. Young, A. R., Bjorn, L. O., Moan, J. & Nultsch, W. (Plenum, New York), pp. 345–377.

26. Blaustein, A. R., Hoffman, P. D., Hokit, D. G., Kiesecker, J. M., Walls, S. C. & Hays, J. B. (1994) *Proc. Natl. Acad. Sci. USA* 91, 1791–1795.

27. Carlini, D. B. & Regan, J. D. (1995) *J. Exp. Mar. Biol. Ecol.* 219, 219–232.

28. Blakefield, M. I. & Harris, D. O. (1994) *Photochem. Photobiol.* 59, 204–208.

29. Ryan, K. G. (1992) *J. Photochem. Photobiol.* B 13, 235–240.

30. Weiler, C. S. & Penhale, P. A., eds. (1994) *Ultraviolet Radiation in Antarctica: Measurements and Biological Effects,* Antarctic Research Series (Am. Geophys. Union, Washington, DC), Vol. 62.

31. Dohler, G. (1992) *Mar. Biol.* (Berlin) 112, 485–490.

32. Karentz, D. (1991) *Antarct. Sci.* 3, 3–11.

33. Roberts, L. (1989) *Science* 244, 288–289.

34. Bothwell, M. L., Sherbot, D. M. J. & Pollock, C. M. (1994) *Science* 265, 97–100.

35. Sancar, A. (1994) *Science* 266, 1954–1957.

36. Kellerman, A. (1989) in *Catalogue of Early Life History Stages of Antarctic Notothenioid Fishes*, BIOMASS Science Series No. 10, ed. Kellerman, A. (Alfred-Wegener-Institut, Bremerhaven, Germany), pp. 45–136.

37. Siegel, V. (1992) *Arch. Fischereiwiss.* 41, 101–130.

38. Hunter, J. R., Taylor, J. H. & Moser, H. G. (1979) *Photochem. Photobiol.* 29, 325–338.

39. Sambrook, J., Fritsch, E. F. & Maniatis, T., eds. (1989) *Molecular Cloning* (Cold Spring Harbor Lab. Press, Plainview, NY), Vol. 2, pp. 9.16–9.19.

40. Mitchell, D. L, Clarkson, J. M., Chao, C. C.-K. & Rosenstein, B. S. (1986) *Photochem. Photobiol.* 43, 595–597.

41. Kneib, R. T. (1986) *Am. Zool.* 26, 259–269.

42. Hardy, J. D., Jr. (1978) *Development of Fishes of the Mid-Atlantic Bight: An Atlas of Egg, Larval, and Juvenile Stages.* (U.S. Fish & Wildlife Service, Fort Collins, CO), USFWS Publ. No. FWSy OBS-78112.

43. Mitchell, D. L., Scoggins, J. T. & Morizot, D. C. (1993) *Photochem. Photobiol.* 58, 455–459.

44. Kellerman, A. & Kock, K.-H. (1989) in *Antarctic Ocean and Resources Variability*, ed. Sahrhage, D. (Springer, Berlin), pp. 147–159.

45. Yasuihira, S., Mitani, H. & Shima, A. (1992) *Photochem. Photobiol.* 55, 97–101.

46. Karentz, D., Cleaver, J. E. & Mitchell, D. L. (1991) *J. Phycol.* 27, 326–341.

47. Mitchell, D. L. & Hartman, P. S. (1990) *BioEssays* 12, 74–79.

48. El-Sayed, S., ed. (1994) *Southern Ocean Ecology* (Cambridge Univ. Press, Cambridge, U.K.).

49. North, A. W. (1991) in *Biology of Antarctic Fish*, eds. di Prisco, G., Maresca, B. & Tota, B. (Springer, New York), pp. 70–86.

50. Morales-Nin, B., Palomera, I. & Schadwinkel, S. (1995) *Polar Biol.* 15, 143–154.
51. Targett, T. E. (1981) *Mar. Ecol. Prog.* Ser. 4, 243.
52. Wormuth, J. H. (1990) *Polar Biol.* 13, 171–182.
53. Bidigare, R. R. (1989) *Photochem. Photobiol.* 50, 469–477.
54. Hubolt, G. (1991) in *Biology of Antarctic Fish*, eds. di Prisco, G., Maresca, B. & Tota, B. (Springer, New York), pp. 1–22.

Reprinted with permission from *Proceedings of the National Academy of Sciences*, Vol. 94, February 1997, pp. 1258-1263. © 1997, National Academy of Sciences, U.S.A.

Earth has a natural sunscreen for UVB radiation. This protective filter is called the ozone layer. Ozone is a form of oxygen that resembles a blue-tinged gas. It forms naturally in the atmosphere. About twenty to thirty miles (thirty-two to forty-eight kilometers) above us lies a thick mat of ozone that blocks out most damaging solar ultraviolet radiation.

If a person slathers on sunblock on a hot, sunny day and misses a spot, that area likely will burn. Similarly, if an area is not covered by ozone, unprotected animals and plants can suffer UVB radiation damage. Human-produced chemicals, such as old refrigerants, have been linked to ozone depletion. This depletion particularly merits concern during the springtime, due to extended sunny days, and in the Antarctic, where UVB rays can be more pronounced. As the authors of the following article reveal, photosynthesis in plants

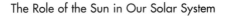

can be negatively affected by UVB rays. Photosynthesis is the biochemical process plants use to turn light energy into food energy.

Just as animals like zooplankton appear to use their own energy to fight off the UVB invasion, certain plants native to the Antarctic Peninsula sometimes grow thicker leaves in response to the rays. Exposure to UVB radiation appears to reduce photosynthesis rates, which could impair the growth rate of the plants. Since plants are at the bottom of Antarctic ecosystems, it can be surmised that insects and animals that live or feed upon the plants also suffer adverse impacts from the UVB exposure. —JV

From "Effect of Solar Ultraviolet-B Radiation During Springtime Ozone Depletion on Photosynthesis and Biomass Production of Antarctic Vascular Plants"
by Fusheng S. Xiong and Thomas A. Day
Plant Physiology, February 2001

Introduction

Increases in solar UV-B radiation (280–315 nm) reaching the Earth's surface due to stratospheric ozone depletion (Madronich et al., 1998) have raised concerns about UV-B impacts on plants (Caldwell et al., 1998). The influence of UV-B on Antarctic organisms is of particular relevance since ozone depletion and corresponding enhancements in solar UV-B are most pronounced in Antarctica (Madronich et al., 1998). For example, ozone concentrations over Antarctica can decline by one-half during austral spring and lead to a doubling in levels of solar UV-B (Frederick and Lubin, 1994; Booth et al., 1998a).

Few studies have examined the influence of solar UV-B on Antarctic biota, and the vast majority of these have focused on marine phytoplankton; solar UV-B levels in Antarctica can depress photosynthesis in these microorganisms, resulting in reductions in marine productivity of 5% to 20% (Smith et al., 1992; Prézelin et al., 1994). Few studies have examined the influence of UV-B on Antarctic terrestrial plants. Regarding the two vascular plant species native to Antarctica (*Colobanthus quitensis* [Kunth] Bartl. and *Deschampsia antarctica Desv.*), Day et al. (1999) and Ruhland and Day (2000) excluded UV-B over naturally growing plants near Palmer Station and found that ambient UV-B levels reduced the vegetative growth of both species. However, it is unknown whether ambient UV-B levels also reduce photosynthesis in these species, and whether growth reductions are correlated with reductions in photosynthesis.

Generalizations about how ecologically realistic UV-B levels affect photosynthesis and whether this in turn affects plant growth under field conditions have been tenuous. Plant exposure to UV-B indoors can impair all major processes in photosynthesis including photochemical reactions in thylakoid membranes, enzymatic processes in the Calvin cycle, and stomatal limitations to CO_2 diffusion (Bornman, 1989; Allen et al., 1998). Several studies have shown that photosystem II (PSII) is often sensitive to UV-B and it has often been assumed to

be the most sensitive photosynthetic target for UV-B (Bornman, 1989; Melis et al., 1992). However, UV-B-induced reductions in CO_2 assimilation can occur prior to, or in the absence of, depressions in PSII function and may more likely involve impairments in the Calvin cycle, possibly mediated by Rubisco (Nogués and Baker, 1995; Lesser and Neale, 1996; Allen et al., 1999). Most studies examining the influence of UV-B on photosynthesis have been conducted in growth chambers or greenhouses and the low background levels of UV-A radiation (315–400 nm) and visible or photosynthetically active radiation (PAR; 400–700 nm) in these indoor studies typically exaggerate UV-B responses, compared with those found in more spectrally realistic outdoor studies (Caldwell et al., 1994). Hence, it is unclear not only as to what photosynthetic target is most sensitive to UV-B, but also whether photosynthesis is even responsive to UV-B under ecologically realistic outdoor spectral regimes.

In this study we first examined the relationship between atmospheric ozone content and UV-B levels during spring along the Antarctic Peninsula to determine whether ozone depletion appeared responsible for enhanced levels of UV-B. We also placed UV-B exclusion filters over vascular plants growing along the Peninsula to assess whether they were responsive to solar UV-B. We compared plants growing under UV-B transparent filters with those under UV-B absorbing filters with respect to their: (a) photosynthetic performance, which we assessed by measuring leaf chlorophyll fluorescence parameters and photosynthetic oxygen evolution rates, (b) ability to recover from high PAR and UV-B-induced impairments in photosynthesis, (c) concentrations of leaf soluble UV-B-absorbing compounds, chlorophyll, and carotenoids, and (d) biomass and leaf area production.

Results and Discussion

Treatment Microclimate

Analysis of the microclimatic data from the UV-B treatments revealed that diurnal (PAR > 100 µmol m^{-2} s^{-1}) mean daily UV-B$_{BE}$ (biologically effective UV-B based on Caldwell's [1971] generalized plant damage action spectrum) under near-ambient UV-B (Aclar filter) and reduced UV-B (Mylar filter) treatments averaged 83 % and 13 %, respectively, of ambient levels over the course of the experiment (October 17, 1998–January 10, 1999). Mean daily UV-A irradiance and PAR under both UV-B treatments averaged 80 % and 90 %, respectively, of ambient levels. Mean diurnal and diel canopy air temperatures in both UV-B treatments were elevated approximately 5°C and 3.5°C, respectively, above ambient.

Solar UV-B Is Negatively Correlated with Ozone Column Content

During the experiment there were three periods of relatively severe ozone depletion (130–210 DU [Dobson units]) that occurred in mid-October, early November, and early December (Fig. 1A [in the original article]). The average ozone column content over the experimental period was 281 DU, which translates into a 20 % ozone depletion, assuming an unperturbed ozone column of 350 DU (Lubin et al., 1992; Frederick and Lubin, 1994).

Linear least-squares correlation/regression analyses of ozone concentrations and midday UV-B$_{BE}$ revealed a significant negative correlation between these variables (P < 0.01; r^2 = 0.31; Fig. 1E, inset, [in original article]). Because of possible non-linearity, as well as uncertainties as to whether these data met normality and homoscadasisity assumptions, we also examined this relationship using Spearman's rank correlation analysis (Sokal and Rohlf, 1981). This analysis also showed a significant negative correlation between ozone column content and midday UV-B$_{BE}$ (P < 0.01; r_s = −0.50). To take into account some of the variability imposed on UV-B$_{BE}$ by cloud cover and solar angle, we also examined trends in UV-B$_{BE}$ by using the ratio of UV-B$_{BE}$: PAR. Linear least-squares regression analysis of ozone column content and midday UV-B$_{BE}$: PAR showed an even stronger, significant negative relationship between these variables (P < 0.01, r^2 = 0.68), as did Spearman's rank correlation analysis (P < 0.01; r_s = 0.80; Fig. 1E, [in original article]).

The significant negative correlations between ozone column content and midday levels of UV-B$_{BE}$ (r^2 = 0.31) and UV-B$_{BE}$:PAR (r^2 = 0.68) over the course of the experiment strongly suggest that higher UV-B$_{BE}$ levels and ratios were at least partly attributable to ozone depletion. The ratios of UV-B-to-PAR during our experiment were high compared with values from lower latitudes. For example, the ratio of integrated (not biologically effective) UV-B-to-PAR measured under clear skies at midday in the summer in Logan, UT (Caldwell et al., 1994) and Neuherberg, Germany (Thiel et al., 1996) was 0.0056 and 0.0037, respectively, whereas this ratio averaged 0.0080 (maximum 0.0126) at midday over the course of our experiment.

UV-B Exposure Reduces Biomass and Leaf Area Production

Over the 85-d growth period (October 17, 1998–January 10, 1999), *D. antarctica* and *C. quitensis* plants produced 22% and 11% less total biomass, respectively, under near-ambient UV-B than under reduced UV-B ($P < 0.05$; Fig. 2, A and B [in the original article]). Treatment effects were more pronounced on above ground than below ground biomass production in both species, and *D. antarctica* and *C. quitensis* produced 18% and 11% less above ground biomass, respectively, under near-ambient UV-B ($P < 0.05$). In contrast, there was no significant UV-B treatment effect on root mass production in *C. quitensis* and only a tendency ($P < 0.10$) for less root mass under near-ambient UV-B in *D. antarctica*. Not surprisingly, root-to-shoot ratios tended to be higher under near-ambient UV-B in both species. Cushion diameters of *C. quitensis* plants under near-ambient UV-B were 9% smaller and tillers of *D. antarctica* were 15% shorter than those under reduced UV-B. In addition, *C. quitensis* and *D. antarctica* plants under near-ambient UV-B produced 24% and 31% less total leaf area than those under reduced UV-B (Fig. 2, E and F [in the original article; $P < 0.05$]). Day et al. (1999) and Ruhland and Day (2000) filtered UV-B over naturally growing plants for whole growing seasons (early November–early March) in previous years at other sites near Palmer Station and also found that ambient UV-B reduced vegetative growth of

these species. Taken collectively, these results indicate that solar UV-B along the Peninsula represents an environmental stress that may consistently limit the performance of vascular plants.

UV-B Exposure Increases Specific Leaf Mass (SLM) and Pigment Concentrations

In both species plants exposed to near-ambient UV-B had substantially greater SLM ($P < 0.01$) SLM of *C. quitensis* and *D. antarctica* under near-ambient UV-B was 30% and 25% greater, respectively, than under reduced UV-B ([Fig. 3, A and B, in the original article]). In both species, plants exposed to near-ambient UV-B also had higher leaf concentrations of UV-B absorbing compounds on an area basis ($P < 0.05$; Fig. 4 inset, [in the original article]), and there was a tendency for this trend on a dry-mass basis as well ($P < 0.10$; data not shown). This was true whether we assessed concentrations by measuring absorbance at 300 or 330 nm. Higher concentrations of UV-absorbing compounds were also apparent in the absorbance spectra of methanol extracts that we used to assess photosynthetic pigment concentrations (Fig. 4, [in the original article]). Concentrations of carotenoids on a leaf-areas basis were also significantly higher in both species under near-ambient UV-B (Fig. 4, [in the original article] $P < 0.05$). Total chlorophyll concentrations tended to be higher on a leaf-area basis in both species under near-ambient UV-B, but not on a dry-mass basis (data not shown). There were no significant UV-B treatment effects on the ratio of chlorophyll *a/b* in either species (data not shown).

UV-B Exposure Does Not Affect Leaf Area-Based Photosynthetic Rates

Light-saturated rates of POE (photosynthetic O_2 evolution) on a leaf-area basis were not affected by UV-B treatment in either species on any of the four sampling dates (Fig. 5A [in the original article]). However, on a total chlorophyll-concentration basis, POE was significantly lower under near-ambient UV-B in both species on three of the four sampling dates ($P < 0.05$). Averaging the means from the four sampling dates, POE per chlorophyll concentration was 17% (*C. quitensis*) and 23% (*D. antarctica*) lower in plants under near-ambient UV-B, with individual sampling date means being 8% to 25% lower in *C. quitensis* and 12% to 33% lower in *D. antarctica* under near-ambient UV-B. In addition, POE per leaf dry-mass was also significantly lower under near-ambient UV-B in both species on three of the four sampling dates ($P < 0.05$). Averaging the means from the four sampling dates, POE per dry mass was 21% (*C. quitensis*) and 26% (*D. antarctica*) lower in plants under near-ambient UV-B (Fig. 5C [in the original article]).

To help distinguish UV-B effects on photochemical versus enzymatic reactions of photosynthesis we determined the POE light-response curves on plants collected at midday on two dates. On a leaf-area basis neither the initial slope of the light-response curve nor maximal rates of POE (at high PAR) were affected by UV-B treatment in either species on either date (Fig. 6, A and B [in the original article]). On a chlorophyll basis there were no treatment effects on the initial slopes of the light-response curves in either species (Fig. 6, C and D [in the original article]). However, maximal rates of POE (at high PAR) on a chlorophyll basis were significantly lower in plants of both species under near-ambient UV-B on both sampling dates, in agreement with our previous measurements of light-saturated POE.

Reductions in vegetative growth and biomass production have been detected in other species in ambient UV-B filter-exclusion studies (Ballaré et al., 1996; Krizek et al., 1997, 1998; Mazza et al., 1999). Several mechanisms have been proffered to explain these growth reductions. One candidate is reduced photosynthetic rate per unit leaf area, although there is little evidence for this mechanism in field studies. Although none of the above exclusion studies assessed UV-B effects on photosynthesis, Ballaré et al. (1996) suggested that the reductions in biomass production they observed were not due to impairments in photosynthesis, since the growth analysis parameter net assimilation rate (dry mass produced per leaf area per time) was not affected by UV-B level. In other UV-B studies, reductions in vegetative growth or changes in canopy architecture due to UV-B often occur in the absence of changes in photosynthetic rates per unit leaf area (Beyschlag et al., 1988; Barnes et al., 1990; Adamse and Britz, 1992; Searles et al., 1995; González et al., 1996, 1998; Allen et al., 1998; Hunt and McNeil, 1998), and Fiscus and Booker (1995) and Allen et al. (1998) concluded that photosynthesis in acclimated plants growing outdoors does not appear at risk from UV-B. Consistent with this we found that exposure to substantial, natural increases in UV-B had no effect on leaf area-based rates of POE.

Although there appears to be little evidence that exposure to UV-B outdoors can reduce rates of POE or CO_2 uptake on a leaf-area basis, we did detect reductions in rates of POE on a chlorophyll and dry-mass basis in plants growing under near-ambient UV-B. These plants not only had higher SLM, suggesting that UV-B exposure led to denser, probably thicker, leaves, but also had higher leaf carotenoid concentrations and tended to have higher chlorophyll concentrations. This could explain why POE per leaf area was not affected by UV-B exposure; plants responded to higher UV-B levels by producing thicker leaves that contained more photosynthetic pigments per area, thereby maintaining photosynthetic gas-exchange rates on an area basis. Higher SLM and/or thicker leaves, along with higher concentrations of soluble UV-B–absorbing compounds, have been found in other species in response to UV-B (Day and Vogelmann, 1995; Johanson et al., 1995; Searles et al., 1995; Ballaré et al., 1996). Both responses could reduce damage to targets in the mesophyll by attenuating and increasing the pathlength of UV-B, thereby reducing fluxes in the mesophyll. The costs associated with producing thicker leaves containing more photosynthetic and UV-B–absorbing pigments per unit leaf area are difficult to quantify. However, this would certainly involve allocating more resources to the construction of new leaf area. Over the course of the growing season, the additional resources required for construction of new photosynthetic leaf area due to UV-B exposure could be impressive, and the limitation this imposes on production of new leaf area and subsequent whole-plant

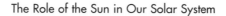
photosynthesis may ultimately explain the reductions in vegetative growth and biomass production we found attributable to solar UV-B.

UV-B Effects on Photosynthesis May Be Associated with Enzymatic Rather than PSII Limitations

Plants of both species under near-ambient UV-B had significantly lower midday Φ_{PSII} (quantum yield of PSII) than those under reduced UV-B on three of the four sunny sampling dates ($P < 0.05$). When we averaged means from all four sampling dates, midday Φ_{PSII} was 8 % (*C. quitensis*) and 16 % (*D. antarctica*) lower in plants under near-ambient UV-B (Fig. 5D [in the original article]). These reductions in Φ_{PSII} under near-ambient UV-B over the four sampling dates ranged from 4 % to 14 % in *C. quitensis* and 8 % to 21 % in *D. antarctica*. The ratio of variable to maximal fluorescence (F_v/F_m) was not affected by UV-B treatment in *C. quitensis* on any sampling date (Fig. 5E [in the original article]). Values were lower in *D. antarctica* under near-ambient UV-B on one of the four sampling dates ($P < 0.05$), but when the means from all four sampling dates were averaged, there was no significant treatment effect.

UV-B treatment also altered the patterns of chlorophyll fluorescence induction curves in both species. F_m was lower and the time taken for fluorescence yield to reach one-half of maximum ($t_{1/2}$) was faster in plants exposed to near-ambient UV-B on all three sampling dates (Fig. 7 [in the original article]). Averaging the means of the three sampling dates under near-ambient UV-B, F_m, and $t_{1/2}$ were 46 % (range 35 %–58 %) and 27 %

(19 % –32 %) lower, respectively, in *C. quitensis*, and 44 % (range 29 % –53 %) and 27 % (17 % –38 %) lower, respectively, in *D. antarctica*. In addition, plants of both species exhibited substantially lower M-peaks under near-ambient UV-B (Fig. 7 [in the original article]). The faster $t_{1/2}$ in both species suggests a smaller plastoquinone pool (Anderson et al., 1988) in our UV-B–exposed plants. A similar shortening of $t_{1/2}$ was observed in diatoms following exposure to enhanced UV-B (Nilawati et al., 1997). Pfündel et al. (1992) reported that violaxanthin deepoxidation was inhibited when plants were exposed to enhanced UV-B, which might increase risks for PAR-induced photoinhibition. Exposure to UV-B may have led to a smaller plastoquinone pool, which may have promoted photoinhibition via over-oxidation of electron carriers around PSII, especially under high PAR.

Although we did not detect any reductions in POE on a leaf-area basis in plants exposed to near-ambient UV-B, we suspect that much of this apparent photosynthetic insensitivity to UV-B may be the result of increases in leaf thickness and chlorophyll concentrations that mitigated any reductions in leaf-area based photosynthetic gas-exchange rates. We did detect reductions in some chlorophyll fluorescence parameters and these data provide information on how UV-B exposure may have affected the photosynthetic apparatus, at least in the upper mesophyll of leaves. Although midday Φ_{PSII} was lower in *C. quitensis* and *D. antarctica* plants under near-ambient UV-B, midday F_v/F_m was unaffected by UV-B treatment. This greater sensitivity of Φ_{PSII} than F_v/F_m to UV-B has previously been observed (Figueroa et al., 1997;

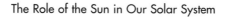

Levall and Bornman, 2000). Nogués and Baker (1995) found that supplemental UV-B lowered the light-saturated CO_2 assimilation rate in the absence of any significant impairments in F_v/F_m in POE. Lesser and Neale (1996) found that although there were no significant differences in F_v/F_m between UV-B exposed and UV-B filtered Antarctic diatoms, the concentrations of large subunits of Rubisco were 20 % lower in the former, and appeared well correlated with a 22 % reduction in CO_2 assimilation rates. Consistent with this, we found that UV-B exposure led to lower Φ_{PSII}, as well as POE per unit chlorophyll and dry mass, but had no affect on F_v/F_m or the initial slope of the light response curves, suggesting that impairment of photosynthesis was associated with light-independent enzymatic limitations, rather than structural damage or photochemical dysfunction of Φ_{PSII}.

A corollary to this idea is that most of the chlorophyll fluorescence signal emitted from the leaf surface originates in the outer 50 µm of leaves (Bornman et al., 1991). Hence, we suspect that leaf surface chlorophyll fluorescence signals may overestimate reductions in whole-leaf photosynthetic rates because they focus on the status of the surface layers of the mesophyll and do not take into account the status of photosynthetic machinery deeper in the mesophyll (Day and Vogelmann, 1995). This bias could be particularly evident in the case of UV-B because damage would likely be most pronounced in surface layers of the mesophyll, and leaves tend to thicken with UV-B exposure such that the contribution of deeper layers of the mesophyll to whole-leaf photosynthesis would probably increase (Day and Vogelmann, 1995), but would go undetected with surface fluorescence measurements. This may explain the discrepancies between rates of leaf-area based POE, which were unaffected by UV-B exposure, and rates of Φ_{PSII}, which were consistently reduced by UV-B.

Φ_{PSII} and F_v/F_m Appear More Sensitive to PAR than UV-B (BH)

Both species showed similar diurnal patterns in Φ_{PSII}, which were characterized by a midday depression and recovery beginning in late afternoon. These midday depressions were evident for both species under both UV-B treatments on both sunny and cloudy days. Figure 8 [in the original article] shows diurnal patterns for a sunny (November 19) and cloudy (November 21) day, and are representative for patterns on the other two pairs of sunny/cloudy days. Several points are apparent from these patterns. First, the midday reductions in Φ_{PSII} tended to be more pronounced under near-ambient UV-B than under reduced UV-B, particularly in *D. antarctica*. On all six sampling dates, Φ_{PSII} in *D. antarctica* was significantly lower at midday (1 PM; $P < 0.05$) and tended to be significantly lower in mid-afternoon (4 PM; $P < 0.10$) under near-ambient UV-B than under reduced UV-B. In *C. quitensis*, Φ_{PSII} tended to be lower ($P < 0.10$) at midday under near-ambient UV-B on two of the six sampling dates. Although UV-B treatments had a significant effect on midday Φ_{PSII}, at least in *D. antarctica*, the depressions in midday Φ_{PSII} appeared much more attributable to the PAR/UV-A wavebands than UV-B. For example, averaging across all six sampling days, we found that from early

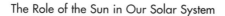

morning (8 AM) to midday (1 PM), Φ_{PSII} in *D. antarctica* dropped by 21 % in plants under reduced UV-B. This reduction in Φ_{PSII} was 28 % in plants under near-ambient UV-B, suggesting that the addition of UV-B contributed to a further reduction in Φ_{PSII} of only 7 %, on average. This corresponds to 25 % (7/28) of the midday depression in Φ_{PSII} being attributable to UV-B. In a similar manner, in *C. quitensis* the midday depression in FPSII was 21 % in plants under reduced UV-B and increased to 25 % in plants under near-ambient UV-B, suggesting that 16 % (4/25) of the midday depression in Φ_{PSII} was attributable to UV-B. Also, the midday depressions were more pronounced in both species on sunny than on cloudy days, implying that high irradiance was at least partly responsible for these depressions in Φ_{PSII}. For example, midday Φ_{PSII} in *D. antarctica* averaged only 0.51 on the three sunny days, compared with 0.66 on the three cloudy days. In *C. quitensis*, midday Φ_{PSII} averaged 0.55 on the sunny days compared with 0.64 on the cloudy days. Last, both species showed relatively fast recovery from these midday depressions under outdoor conditions. Averaging the means from all six sampling dates we found that on average, Φ_{PSII} had recovered 74 % (*C. quitensis*) and 67 % (*D. antarctica*) of its midday depression by 7:30 PM.

Although there were no significant UV-B treatment effects on midday F_v/F_m in either species, our diurnal measurements confirmed that midday values were depressed in both species, and these depressions were more pronounced on sunny than cloudy days. For example, on the three sunny sampling dates, mean midday F_v/F_m in *D. antarctica* and *C. quitensis* declined by 16 %

and 14%, respectively, of their early morning values, whereas on cloudy days, midday F_v/F_m in *D. antarctica* and *C. quitensis* declined by 8% and 6%, respectively, of their early morning values (diurnal data not shown).

Our findings support the idea that PSII is more sensitive to high visible (or UV-A) irradiance than UV-B (Allen et al., 1999). Only 16% to 25% of the midday depressions we observed in Φ_{PSII} appeared attributable to UV-B, and furthermore, the midday depressions in F_v/F_m were not affected by UV-B exposure. Krause et al. (1999) found that exposure of two tropical species to sunlight resulted in substantial midday depressions in F_v/F_m, but these depressions were still very evident when UV-B was excluded, and we estimate that < 15% of the reductions they observed were attributable to UV-B. In a similar manner, only about 5% to 12% of the midday depressions in photosynthesis in marine algae appear attributable to UV-B (Figueroa et al., 1997; Herrmann et al., 1997; Gómez et al., 1998).

Appreciable Recovery of Φ_{PSII} at Low Temperatures

Because low temperatures can impede the recovery from photoinhibition following exposure to high irradiance, at least in temperate and tropical species (Gong and Nilsen, 1989; Sukhvibul et al., 2000), we assessed the effect of temperature on recovery of Φ_{PSII} by removing plants from midday sunlight and placing them in incubators under low visible irradiance at a temperature of 4°C or 12°C. In both species, recovery from midday depression of Φ_{PSII} was faster in plants at 12°C than 4°C (Fig. 9 [in the original article]). At the higher temperature, Φ_{PSII} had

recovered 86 % (*C. quitensis*) and 81 % (*D. antarctica*) of its depression from early morning (8 AM) values after 8 h indoors. However, recovery was appreciable even at the lower temperature, and Φ_{PSII} had recovered 60 % (*C. quitensis*) and 55 % (*D. antarctica*) of its early morning values after 8 h at 4°C, which is impressive considering that recovery in temperate and tropical species is much slower or eliminated at 3°C to 8°C (Gong and Nilsen, 1989; Sukhvibul et al., 2000).

Were Plants Responsive to Day-to-Day Variations in Solar UV-B Levels?

A relevant, although rarely tested, question is whether plants are responsive to variations in ambient UV-B levels from day to day. Some plant responses such as DNA damage (Stapleton and Walbot, 1994; Kang et al., 1998), D1 protein degradation (Jansen et al., 1996), and anthocyanin synthesis (Hada et al., 1996) can be approximately linearly related to UV-B dose in short-term laboratory experiments. However, whether plant responses are significantly correlated with fluctuations in natural UV-B levels outdoors over periods of several days or weeks has rarely been tested. Ballaré et al. (1996) found that transmission of greater percentages of ambient UV-B led to corresponding increases in leaf DNA damage levels in a summer annual. In one of the few studies to examine the relationship between natural temporal fluctuations in UV-B levels and plant response Rousseaux et al. (1999) found that fluctuations in ambient $UV-B_{BE}$ dose explained a large proportion (68 %) of the variation in leaf DNA damage levels over 14 sampling dates during springtime in southern Argentina. We found no significant correlations between any of the photosynthetic variables we measured and several UV-B parameters we examined, including midday UV-B$_{BE}$, midday UV-B$_{BE}$-to-PAR, and daily UV-B$_{BE}$ dose. The largest photosynthetic data set we had for this analysis was midday Φ_{PSII} (10 d) and regardless of whether we expressed this in terms of percentage of inhibition (near-ambient/reduced UV-B treatment) or absolute Φ_{PSII} rates under near-ambient UV-B, the relationships were weak ($P > 0.20$; $r^2 < 0.25$). Examination of these correlations with a non-linear test (Spearman's rank correlation) also failed to detect any significant correlations. The lack of correlation between UV-B level and photosynthetic inhibition could be due to several factors, including: (a) Inhibition was saturated by relatively low levels of ambient UV-B, (b) response to other environmental factors such as air temperature (Xiong et al., 1999, 2000) may have confounded or overshadowed their photosynthetic responses to UV-B, (c) acclimation or protective responses might have occurred over the season, as well as during short periods of high UV-B levels, or simply (d) the relatively small sample size of our data set.

Conclusions

We provide evidence that ozone depletion was at least partly responsible for enhanced levels of UV-B along the Antarctic Peninsula during this experiment. Furthermore, exposure of native vascular plants to these UV-B levels led to appreciable reductions in biomass production and leaf area. Rates of photosynthetic gas exchange, on a leaf area basis, were not affected by exposure to UV-B, and cannot explain these reductions in growth. Leaves on plants exposed to UV-B were denser, probably thicker, and had higher concentrations of photosynthetic and UV-B-absorbing pigments. We suspect that the development of thicker leaves containing more photosynthetic pigments allowed these plants to maintain their photosynthetic rates per unit leaf area at rates similar to plants under reduced UV-B levels. However, the additional resources required for construction of leaf area, and subsequent reductions in whole-plant photosynthetic surface area over the course of the season might ultimately explain the reductions in growth and biomass reductions we found attributable to solar UV-B. Exposure to solar UV-B did reduce Φ_{PSII}, although F_v/F_m was unaffected, suggesting that UV-B did impair photosynthesis, at least in the upper mesophyll of leaves, and that this was associated with light-independent enzymatic limitations, rather than direct damage to PSII.

Literature Cited

Adamse P, Britz SJ (1992). Amelioration of UV-B damage under high irradiance: I. Role of photosynthesis. Photochem Photobiol 56: 645–654

Allen DJ, Nogués S, Baker NR (1998). Ozone depletion and increased UV-B radiation: is there a real threat to photosynthesis? J Exp Bot 49: 1775–1788

Allen DJ, Nogués S, Morison JIL, Greenslad PD, McLeod AR, Baker NR (1999). A thirty percent increase in UV-B has no impact on photosynthesis in well-watered and droughted pea plants in the field. *Global Change Biol 5*: 235–244

Anderson JM, Chow WS, Goodchild DJ (1988). Thylakoid membrane organization in sun/shade acclimation. In Evans JR, Von Caemmerer S, Adams WW, III, eds, Ecology of Photosynthesis in Sun and Shade. Commonwealth Scientific and Industrial Research Organization, Melbourne, Australia, pp 11–16

Ballaré CL, Scopel AL, Stapleton AE, Yanovsky MJ (1996). Solar ultraviolet-B radiation affects seedling emergence, DNA integrity, plant morphology, growth rate, and attractiveness to herbivore insects in Datura ferox. Plant Physiol 112: 161–170

Barnes PW, Flint SD, Caldwell MM (1990). Morphological responses of crop and weed species of different growth forms to ultraviolet-B radiation. Am J Bot 77: 1354–1360

Beyschlag W, Barnes PW, Flint SD, Caldwell MM (1988). Enhanced UV-B irradiation has no effect on photosynthetic characteristics of wheat (*Triticum aestivum L.*) and wild oat (*Avena fatua L.*) under greenhouse and field conditions. Photosynthetica 22: 516–525

Booth CR, Diaz SB, Cabasug LW, Mestechkina T, Robertson JS, Ehramjian JC (1998a). UV irradiance measurements in Antarctica: results of ten years of data collection. Abstracts of the European Conference on Atmospheric UV Radiation. Finnish Meteorological Institute, Helsinki, Finland, pp. 9

Booth CR, Ehramjian JC, Mestechkina T, Cabasug LW, Robertson JS, Tusson IVJR (1998b). NSF Polar Programs UV Spectroradiometer Network, 1995–1997 Operations Report. Biospherical Instruments, San Diego

Bornman JF (1989). Target sites of UV-B radiation in photosynthesis of higher plants. J Photochem Photobiol B Biol 4: 145–156

Bornman JF, Vogelmann TC, Martin G (1991). Measurement of chlorophyll fluorescence within leaves using a fiber optic microprobe. Plant Cell Environ 14: 719–725

Caldwell MM (1971). Solar UV irradiation and the growth and development of higher plants. In Giese AC, ed, Photophysiology Academic Press, New York, pp. 131–177

Caldwell MM, Flint SD, Searles PS (1994). Spectral balance and UV-B sensitivity of soybean: a field experiment. Plant Cell Environ 17: 267–276

Caldwell MM, Teramura AH, Tevini M, Bornman JF, Björn LO, Kulandaivelu G (1998). Effects of increased solar ultraviolet radiation on terrestrial ecosystems. J Photochem Photobiol B Biol 46: 40–52

Day TA, Ruhland CT, Grobe CW, Xiong FS (1999). Growth and reproduction of Antarctic vascular plants in response to warming and UV radiation reductions in the field. Oecologia 119: 24–35

Day TA, Vogelmann TC (1995). Alterations in photosynthesis and pigment distributions in pea leaves following UV-B exposure. Physiol Plant 94: 433–440

Figueroa FL, Salles S, Aguilera J, Jiménez C, Mercado J, Viñegla B, Flores-Moya A, Altamirano M (1997). Effect of solar radiation on photoinhibition and pigmentation in the red alga *Porphyra leucosticta*. Mar Ecol Prog Ser 151: 81–90

Fiscus EL, Booker FL (1995). Is increased UV-B a threat to crop photosynthesis and productivity? Photosynth Res 43: 81–92

Frederick JE, Lubin D (1994). Solar ultraviolet irradiance at Palmer Station, Antarctica. *In* Weiler CS, Penhale PA, eds, Ultraviolet Radiation in Antarctica: Measurements and Biological Effects. Antarctic Research Series American Geophysical Union, Washington, DC, pp. 43–52

Genty B, Briantais MJ, Baker NR (1989). The relationship between the quantum yield of photosynthetic electron transport and quenching of chlorophyll fluorescence. Biochim Biophys Acta 990: 87–92

Gómez I, Pérez-Rodríguez E, Viñegla B, Figueroa FL, Karsten U (1998). Effects of solar radiation on photosynthesis, UV-absorbing compounds and enzyme activities of the green alga *Dasycladus vermicularis* from southern Spain. J Photochem Photobiol B Biol 47: 46–57

Gong H, Nilsen S (1989). Effect of temperature on photoinhibition of photosynthesis, recovery, and turnover of the 32 kD chloroplast protein in *Lemna gibba*. J Plant Physiol 135: 9–14

González R, Mepsted R, Wellburn AR, Paul ND (1998). Non-photosynthetic mechanisms of growth reduction in pea (*Pisum sativum L.*) exposed to UV-B radiation. Plant Cell Environ 21: 23–32

González R, Paul ND, Percy K, Ambrose M, Mclaughlin CK, Barnes JD, Areses M, Wellburn AR (1996). Responses to ultraviolet-B radiation (280–315 nm) of pea (*Pisum sativum*) lines differing in leaf surface wax. Physiol Plant 98: 852–860

Hada M, Tsurumi S, Suzuki M, Wellmann E, Hashimoto T (1996). Involvement and non-involvement of pyrimidine dimer formation in UV-B effects on Sorghum bicolor Moench seedlings. J Plant Physiol 148: 92–99

Herrmann H, Häder D-P, Ghetti F (1997). Inhibition of photosynthesis by solar radiation in *Dunaliella salina*: relative efficiencies of UV-B, UV-A and PAR. Plant Cell Environ 20: 359–365

Hunt JE, McNeil DL (1998). Nitrogen status affects UV-B sensitivity of cucumber. Aust J Plant Physiol 25: 79–86.

Jansen MAK, Gaba V, Greenberg BM, Mattoo AK, Edelman M (1996). Low threshold levels of ultraviolet-B in a background of photosynthetically active radiation trigger rapid degradation of the D2 protein of photosystem II. Plant J 9: 693–699.

Johanson U, Gehrke C, Björn LO, Callaghan TV (1995). The effects of enhanced UV-B radiation on the growth of dwarf shrubs in a subarctic heathland. Funct Ecol 9: 713–719

Kang HS, Hidema J, Kumagai T (1998). Effects of light environment during culture on UV-induced cyclobutyl pyrimidine dimers and their photorepair in rice (Oryza sativa L.). Photochem Photobiol 68: 7177

Krause GH, Schmude C, Garden H, Koroleva OY, Winter K (1999). Effects of solar ultraviolet radiation on the potential efficiency of photosystem II in leaves of tropical plants. Plant Physiol 121: 1349–1358.

Krizek DT, Britz SJ, Mirecki RM (1998). Inhibitory effects of ambient levels of solar UV-A and UV-B radiation on growth of cv. New Red Fire lettuce. Physiol Plant 103: 17.

Krizek DT, Mirecki RM, Britz SJ (1997). Inhibitory effects of ambient levels of solar UV-A and UV-B radiation on growth of cucumber. Physiol Plant 100: 886–893.

Lesser MP, Neale PJ (1996). Acclimation of Antarctic phytoplankton to ultraviolet radiation: ultraviolet-absorbing compounds and carbon fixation. Mol Marine Biol Biotechnol 5: 314–325

Levall MW, Bornman JF (2000). Differential response of a sensitive and tolerant sugar-beet line to *Cercospora beticola* infection and UV-B radiation. Physiol Plant 109: 21–27

Lichtenthaler HT, Wellburn AR (1983). Determination of total carotenoid and chlorophyll a and b of leaf segments in different species. Biochem Soc Trans 11: 591–592

Lubin D, Mitchell BG, Frederick JE, Albert AD, Booth CR, Lucas T, Neuschuler D (1992). A contribution toward understanding the biospherical significance of Antarctic ozone depletion. J Geophys Res 97: 7817–7828.

Madronich S, McKenzie RL, Björn LO, Caldwell MM (1998). Changes in biologically active ultraviolet radiation reaching the Earth's surface. J Photochem Photobiol B Biol 46: 519.

Mazza CA, Battista D, Zima AM, Szwarcberg-Bracchitta M, Giordana CV, Acevedo A, Scopel AL, Ballaré CL (1999). The effects of solar ultraviolet-B radiation on the growth and yield of barley are accompanied by increased DNA damage and antioxidant responses. Plant Cell Environ 22: 61–70.

Melis A, Nemson JA, Harrison MA (1992). Damage to functional components and partial degradation of photosystem II reaction center proteins upon chloroplast exposure to ultraviolet-B radiation. Biochim Biophys Acta 1100: 312–320

Nilawati J, Greenberg BM, Smith REH (1997). Influence of ultraviolet radiation on growth and photosynthesis of two cold ocean diatoms. J Phycol 33: 215–224.

Nogués S, Baker NR (1995). Evaluation of the role of damage to photosystem II in the inhibition of CO_2 assimilation in pea leaves on exposure to UV-B radiation. Plant Cell Environ 18: 781–787.

Pfündel EE, Pan RS, Dilly RA (1992). Inhibition of violaxanthin deepoxidation by ultraviolet-B radiation in isolated chloroplasts and intact leaves. Plant Physiol 98: 1372–1380.

Prézelin BB, Boucher NB, Smith RC (1994). Marine primary production under the influence of the Antarctic ozone hole: Icecolors '90. In Weiler CS, Penhale PA, eds, Ultraviolet Radiation in Antarctica: Measurements and Biological Effects. Antarctic Research Series American Geophysical Union, Washington, DC, pp 159–186.

Rousseaux MC, Ballaré C, Giordano CV, Scopel AL, Zima AM, Szwarcberg-Bracchitta M, Searles PS, Caldwell MM, Diaz SB (1999). Ozone depletion and UVB radiation: impact on plant DNA damage in southern South America. Proc Natl Acad Sci USA 96:15310–15315.

Ruhland CT, Day TA (2000). Effects of ultraviolet-B radiation on leaf elongation, production and phenylpropanoid concentrations of *Deschampsia antarctica* and *Colobanthus quitensis* in Antarctica. Physiol Plant 109: 244–251.

Schreiber U, Schiwa U, Bilger W (1986). Continuous recording of photochemical and non-photochemical chlorophyll fluorescence quenching with a new type of modulation fluorometer. Photosynth Res 10: 51–62.

Searles PS, Caldwell MM, Winter K (1995). The response of five tropical dicotyledon species to solar ultraviolet-B radiation. Am J Bot 82: 445–453.

Sivak MN, Walker DA (1985). Chlorophyll *a* fluorescence: can it shed light on fundamental questions in photosynthetic carbon dioxide fixation? Plant Cell Environ 8: 439–448.

Smith RC, Prézelin BB, Baker KS, Bidigare RR, Boucher NP, Coley T, Karentz D, MacIntyre S, Matlick HA, Menzies D, Ondrusek M, Wan Z, Waters KJ (1992). Ozone depletion: ultraviolet radiation and phytoplankton biology in antarctic waters. Science 255: 952–959.

Smith RC, Stammerjohn SE, Baker KS (1996). Surface air temperature variations in the western Antarctic Peninsula region. *In* Ross RM, Hofman EE, Quetin LB, eds, Foundations for Ecological Research West of the Antarctic Peninsula. Antarctic Research Series American Geophysical Union, Washington, DC, pp. 105–121.

Sokal PR, Rohlf FJ (1981). Biometry. WH Freeman and Company, New York

Stapleton AE, Walbot V (1994). Flavonoids can protect maize DNA from the induction of ultraviolet radiation damage. Plant Physiol 105: 881–889.

Sukhvibul N, Whiley AW, Smith MK, Hetherington SE (2000). Susceptibility of mango (*Mangifera indica L.*) to cold-induced photoinhibition and recovery at different temperatures. Aust J Agric Res 51: 503–513.

Thiel S, Döhring T, Köfferlein M, Kosak A, Martin P, Seidlitz HK (1996). A phytotron for plant stress research: how far can artificial lighting compare to natural sunlight? J Plant Physiol 148: 456–463.

Xiong FS, Mueller EC, Day TA (2000). Photosynthetic and respiratory acclimation and growth response of Antarctic vascular plants to contrasting temperature regimes. Am J Bot 87: 700–710.

Xiong FS, Ruhland CT, Day TA (1999). Photosynthetic temperature response of the Antarctic vascular plants *Colobanthus quitensis* and Deschampsia antarctica. Physiol Plant 106: 276–286.

While this next paper, "Perception of Solar UVB Radiation by Phytophagous Insects: Behavioral Responses and Ecosystem Implications," looks at a soybean crop system in Buenos Aires, Argentina, the research is an indirect extension of the work described in the previous article. That is

because the scientists here look at not only how UVB radiation affects plants, which in this case are soybeans, but also how the soybean reaction to the UVB exposure impacts bugs that feed from the plants. The bugs here are thrips, which are minute sucking insects that ingest plant juices.

The researchers found that thrips preferred to feed on soybeans that were not exposed to solar ultraviolet radiation. That is evidence that the bugs sensed something was wrong with those plants. The bugs also appeared to sense UVB rays and would try to avoid them at all cost. Additional experiments on soybean worms revealed that the worms would avoid consuming leaves that had been sucked on by thrips. While one might conclude that selective use of UVB rays could prohibit some insect damage to crops, this question must be asked: if even bugs avoid UVB-affected plants, should not humans be concerned about such plants, too? —JV

"Perception of Solar UVB Radiation by Phytophagous Insects: Behavioral Responses and Ecosystem Implications"
by Carlos A. Mazza, Jorge Zavala, Ana L. Scopel, and Carlos L. Ballaré
Proceedings of the National Academy of Sciences,
February 1999

Introduction

Depletion of stratospheric ozone (1) is a cause of concern because the biological impacts of an increase in

solar UVB (290–320 nm) are unknown. Most studies to date on ecological effects of solar UVB have been carried out on plants (2–4); however, there is growing awareness in the UVB research community (2–7), as well as among those studying other aspects of global environmental change (8), about the limitations of impact predictions that result from up-scaling information obtained in studies of a single organism or trophic level. The concentration of the research effort on plants is in part a consequence of the assumption that the effects of UVB on ecosystem functioning are largely mediated by its effects on the primary producers (6).

Recent studies on animal consumers have focused on those effects of UVB on consumer growth and survival that are mediated by changes in host chemistry (9–11) and on direct damaging effects of UVB. In the latter regard, it has been shown that certain animal species (particular instars), such as juvenile forms of some aquatic organisms, are not well protected from UV radiation and are damaged by prolonged exposure to present-day levels of UVA (320–400 nm) and UVB (4, 12, 13) radiation. Damaging effects of acute exposures on zooplanktonic organisms have also been documented in studies carried out in Antarctic waters under ozone-hole episodes (14). Interestingly, one of the few studies that included more than one trophic level has shown that direct damaging effects of solar UV radiation on phytophagous insect larvae can counterbalance the negative impact of UV on algal photosynthesis and result in increased biomass accumulation of the primary producers in experimental freshwater ecosystems (5).

Almost no research has been carried out on the effects of UVB on animal behavior. There is some evidence from field studies for indirect, plant-mediated effects of UVB on insect feeding choices (15), but the generality and underlying mechanisms of these effects have not been established. The possibility of direct behavioral responses of animals to solar UVB has received little attention. Controlled-environment studies with protozoans (16) have suggested that some species may detect and avoid exposure to UVB radiation; however, because light sources that emitted predominantly in the UVB region were used in those experiments, it was impossible to separate potentially specific effects of UVB from simple responses induced by changes in total irradiance. Some animals are believed to use UVA and "human-visible" ($\lambda \geqslant 400$ nm) light as a cue to avoid exposure to harmful UVB radiation (see ref. 5). Others, such as the housefly, are attracted to both UVA and UVB sources in indoor experiments, but the response to lamps that emit predominantly in the UVB is weaker than the response to UVA lamps. Roberts *et al.* (16) concluded that houseflies do not have a genuine spectral preference within the UV region; the reduced behavioral response in the UVB is simply a consequence of the decline in photoreceptor response (as evaluated with electroretinograms) toward the UVB region. Thus, although UV photoreceptors whose sensitivity extends down to wavelengths as short as 320 nm have been identified in the visual systems of several species of animals (ref. 18; reviewed in ref. 19), animals are generally thought to be unable to resolve UVB from other wavelengths in natural sunlight.

We used a simple agroecosystem, in which soybeans were the main primary producers and a few species of insects the consumers, to investigate the impacts of solar UVB on plant–consumer interactions. Our results suggest that (*i*) apart from inducing changes in the plant hosts that strongly affect their attractiveness to thrips (*Caliothrips phaseoli*), solar UVB radiation triggers direct behavioral responses in the insects and (*ii*) thrips induce changes in their plant hosts that, in turn, feed forward to other species of phytophagous insects.

Materials and Methods

Plant Culture. All experiments were carried out during the summer and autumn of 1995–1996 and 1996–1997 at the experimental fields of IFEVA in Buenos Aires (34°35' S, 58°29' W). In all experiments, soybean was grown in field plots (1.2 × 1.2 m; plant density = 60 per m²) that were covered with plastic filters designed to exclude various levels of solar UVB. In the 1995–1996 growing season, the plants were contained in 10-liter pots. There were two planting dates: 5 December, 1995 with four genotypes in each plot: CNS, Essex, Lincoln, and Williams (combined population), and 12 February 1996, using the line PI227687. In the 1996–1997 season, the seeds were planted in rows 12 cm apart; four commercial soybean varieties were planted in all plots (each in a different row): A5308, Williams, Charata-76, and Dekalb 458 (sowing date: 28 February 1997). The plots were watered periodically to maintain the soil near field capacity and were kept free of weeds. The autumn of 1997 was exceptionally warm and allowed continued growth of the crops until mid-May.

UV Manipulation Techniques. All plots were covered before crop seedling emergence with plastic film that transmitted more than 88 % of the photosynthetically active radiation (400–700 nm) and attenuated different regions of the UV band. The UVB-transparent control (SUN) plots were covered with either 0.02-mm thick polyethylene (Rolopac, Buenos Aires) or Aclar (0.04 mm thick; Allied Signal Plastics, Morristown, NJ) films, which transmit more than 80 % throughout the UVB and UVA bands. The -UVB plots were covered with Mylar-D film (0.1 mm thick; DuPont), which blocks essentially all radiation below 310 nm (see spectrum in ref. 15). Intermediate UVB irradiances were obtained by superimposing Mylar strips on sheets of the UV-transparent Rolopac or Aclar films (see ref. 15). The relative UVB irradiances under the various filters were measured at midday with a cosine-corrected UVB detector (SUD/240/W) attached to a IL-1700 research radiometer (International Light, Newburyport, MA). The spectral response of the detector head is centered at 290 nm (half-bandwidth = 20 nm) and its noontime signal is reduced by more than 95 % when covered with a Mylar filter, indicating that the detector is virtually blind in the UVA region. In both seasons there were four true replicates (blocks) of each UVB treatment. Each individual field plot was surrounded by an almost continuous soybean canopy, which greatly reduced the contribution of sidelight, and the filters were kept at a short distance ($\approx$5 cm) from the upper-canopy leaves. Consequently, UVB attenuation at the center of the plots was very effective ($\approx$98 % in the −UVB plots compared with the SUN plots, see Fig. 1 [in the original article]); all leaves used for field and laboratory bioassays were collected from plants located at the center of the plots. The plastic filters might have had some influence on the canopy microclimate compared with a no-filter situation. However, this should not affect the comparisons between the UVB treatments used in our experiments, as previous studies have shown no differences between UVB exclusion treatments in leaf or soil temperature (15).

In the short-term UV-supplementation experiments, we irradiated portions of the field plots with light from UVB bulbs placed 50 cm above the canopy. UVB-313 bulbs (Q-Panel, Cleveland) were covered with either a polyethylene film (transparent to UVC, UVB, and UVA), a cellulose diacetate filter (transparent to UVB and UVA), or a Mylar filter (transparent to UVA). The relative energy output of the unfiltered UVB-313 lamps in the UVC (< 290 nm), UVB, and UVA spectral bands is 1 %, 80 %, and 19 %, respectively. The absolute irradiance provided by the lamps at canopy height measured with the UVB radiometer (λ_{max} = 290 nm) was 1.5 10^{-9} W/cm^2, which represented a 10-fold increase over the normal noontime value under clear-sky conditions in May. Measurements taken with the sensor pointing downward showed that the UVB component in the diffuse light received at the abaxial surface of upper-canopy leaves increased 4.25 ± 0.33 times (P < 0.0001) as a result of UVB supplementation with the lamps. For treatment with UVA (see Fig. 3C [in the original article]), we used UVA TL 40W/05 bulbs (Philips). The relative spectral output of the unfiltered TL/05 lamp in the UVC, UVB, and UVA spectral bands is 0.5 %, 2.5 %, and 97 %, respectively.

Insect Surveys and Feeding Experiments. The plants at our field site were colonized by natural populations of thrips, which were particularly abundant in the abaxial surface of leaves in the upper third of the canopy. The insects scraped the leaf surface and eventually induced chlorosis in the affected areas. Insect counts were made at the beginning of flowering (1996: end of March, genotype PI227687; 1997: end of April, all four cultivars combined) on the youngest fully expanded leaf (15–30 randomly selected plants per plot; four independent blocks per treatment). Leaf damage was assessed in late March 1996 (genotype PI227687) by estimating the fractional area that was damaged by thrips in the youngest fully expanded leaf (11th node; 16 plants per plot; 4 plots per treatment). The thrips lesions could be visually identified as areas in which the leaf surface appeared scraped and the inner tissues were slightly chlorotic.

The effect of solar UVB on plant tissue attractiveness to thrips was tested in field and laboratory "choice" experiments. In both cases, the response variable was the number of insects that landed on leaves with contrasting UVB history at various times from the beginning of the experiment. In the field, trifoliate leaves collected from soybean crops grown with (SUN) or without (−UVB) solar UVB (1997 sowing; all cultivars combined) were placed at the center of a soybean canopy that was covered with a −UVB filter. The petioles of the detached leaves were kept continuously under water in a glass jar during the course of the experiment. There were six leaves in each jar (three from the SUN pretreatment and three from the −UVB pretreatment) with a total of 12 independent replicate jars distributed within the soybean canopy. The experiment was carried out on 7 and 8 May 1997. In the laboratory, young fully expanded leaves from −UVB and SUN plants (Genotype PI227687) were offered to locally collected thrips in 35 × 60 × 15 (height)-cm plastic boxes (40 freshly collected adults and three leaves of each treatment per box). The experiment was carried out in three opportunities during April 1996 (the data were pooled for analysis), and there were 15 independent choice boxes.

To test for antiherbivore responses induced by thrips, fully expanded leaves from SUN plants (1995–1996 sowing) with or without visual symptoms of thrips damage were presented in a laboratory choice experiment to larvae of *Anticarsia gemmatalis* (soybean worm), which is an important late-season soybean pest that feeds on aerial parts of the plant and can cause severe yield losses in commercial crops. The experiments were carried out during March and April of 1996 by using leaves from the combined population and the line PI227687. In all "damaged" leaves, the affected areas (scraped leaf surface; slightly chlorotic mesophyll) covered less than 15 % of the lamina and were evident only on careful inspection of the leaves. The larvae were 15 days old at the beginning of the experiment and had been fed on a standard artificial diet (20). The larvae and the leaves were placed inside 35 × 60 × 15 (height)-cm plastic boxes (5 larvae per box); the amount of tissue consumed was estimated from leaf area measurements (LI-3000, Li-Cor, Lincoln, NE) taken before and 24 h after the beginning of the feeding experiment. The initial leaf area was approximately 100 cm² per box for each treatment (thrips damage level); leaves of both treatments were offered in the same box. To prevent leaf desiccation, the petioles were wrapped in cotton saturated with water; temperature varied between 25 and 30 °C. There were 10 replicate boxes.

Data Analysis. All data were analyzed by using the SAS Version 6.12 statistical package (SAS Institute, Cary, NC). The standard errors reported in the figures were calculated from the error-mean-square values (S^2) of the relevant ANOVA tables as SE $= (S^2/n - 1)^{1/2}$.

Results and Discussion

Effects of Solar UVB on Herbivory. In both seasons, filtering out solar UVB resulted in a 3- to 5-fold increase in the density of the thrips populations that invaded our field crops (−UVB vs SUN plots; Fig. 1a [in the original article]). This was a specific effect of UVB exclusion, because (*i*) the filters had virtually the same transmittance above 320 nm, (*ii*) filtering out the UVB component of sunlight changed total photon flux density between 280 and 700 nm by less than 1%, and (*iii*) the crop microenvironment (e.g., soil and canopy temperature) was not differentially affected by UVB treatments (see ref. 15). There was a clear dose-response relationship between UVB levels and thrips density (Fig. 1A [in the original article]), and the greater abundance of thrips in the −UVB treatment resulted in significantly higher leaf damage. This UVB effect on insect abundance and herbivory is much larger than any previously reported effect of solar UVB on plant growth and gross morphology (refs. 2, 3, 15, and refs. therein). Indeed, the soybean crops used in our experiments displayed pigmentation and morphological responses to solar UVB that were only modest in comparison with the effects of UVB exclusion on their herbivores. Thus, in 1995–1996 we found that solar UVB inhibited stem elongation and promoted the accumulation of methanol-soluble phenolics in some of the cultivars; although these effects were statistically significant, they always involved changes of less than ± 30% of the −UVB mean value. In the 1996–1997 crops, which were planted in late summer (with lower ambient UVB levels), we did not detect significant effects of UVB exclusion on stem elongation, leaf area expansion, and total phenolics in any of the cultivars tested (data not shown from measurements taken between 15 and 32 days after sowing).

Plant-Mediated Effects of Solar UVB on Insect Behavior. To test the possibility that the effects of solar UVB on herbivory were mediated by behavioral responses to UVB-induced changes in the plant host (cf. ref. 15), we carried out two preference tests. In one of them, we collected leaves from soybean plants grown in the SUN and −UVB treatments and placed them within a soybean crop that was covered with a −UVB film and presented a dense thrips infestation. Within a few minutes the leaves were invaded by thrips, and the insects clearly preferred leaves from plants grown in the −UVB treatment over leaves of the SUN plants (Fig. 1A [in the original article]). In complementary experiments carried out in the laboratory, we offered soybean leaves from −UVB and SUN plants to captive specimens of thrips. Insect counts (Fig. 2B [in the original article]) again demonstrated that leaves not exposed to solar UVB were preferred over leaves from SUN plants. Previous laboratory experiments have shown that changes in the plant host induced by UVB can affect the performance (growth and survival) of insect herbivores (refs. 9–11, and literature cited in ref. 7). All of these experiments were "no-choice" assays—i.e., the insects were fed with leaves that were either treated or not treated with UVB. Our data clearly show that exposure to solar UVB causes persistent changes in the leaves that are recognized by thrips and used in their host-selection decisions.

Rapid Effects of Solar UVB on Insect Behavior. Animals of many taxa, including invertebrates (18, 21), fishes (22), birds (23, 24), and rodents (25) present visual sensitivity to UV radiation. However, photoreceptor sensitivity curves always have maxima in the UVA region (above 320 nm) (17–19, 21–23, 25), and it would seem unlikely, therefore, that these photoreceptors may detect variations in natural UVB levels. This is particularly true because, in natural sunlight, the spectral irradiance is at least 10 times greater in the UVA than in the UVB region; therefore, detection of solar UVB would require a high photoreceptor specificity. To investigate rapid behavioral responses of thrips to solar UVB, we performed a filter switch experiment. Filters were removed from the field plots, which were divided in two; each half was covered either with the same original film (i.e., SUN → SUN or −UVB → −UVB) or with a film that created the opposite UVB condition (i.e., SUN → −UVB or −UVB → SUN). Surprisingly, switching the filters induced a rapid migration of the insects within the plots and, a few hours after the switch, the greatest thrips densities were found in the plot halves covered by the −UVB film. Rapid changes in host quality (such as those triggered by mechanical damage in some species; ref. 26) could, at least in principle, explain this response to the filter switch; however, we consider this possibility unlikely for two reasons. First, insect surveys carried out 4 h after the switch already showed a highly significant treatment effect ($P = 0.005$; data not shown). Second, even if the decline in tissue quality caused by the −UVB → SUN switch was very rapid, it seems unlikely that the opposite change (SUN → −UVB) could have caused an almost instantaneous increase in leaf attractiveness. In fact, our own data (Fig. 2A [in the original article]) show that the antiherbivore response induced by solar UVB in soybeans is effective at deterring thrips for at least a few hours after the end of the UVB exposure, which is consistent with observations in other systems showing that antiherbivore defenses can be sustained for long periods after induction (for a recent review, see ref. 27).

To further investigate the apparently direct response of the insects to UVB, we irradiated portions of a field plot with light from UVB bulbs that were covered with either a polyethylene film (transparent to UVC, UVB, and UVA), a cellulose diacetate filter (transparent to UVB and UVA), and a Mylar filter (transparent to UVA). After a few minutes of irradiation, we counted thrips in the leaves located under the lamps. Insect density was not changed by illumination with residual UVA from the UVB-313 lamps (Fig 3B, UVAr [in the original article]); in contrast, illumination with UVB or UVB + residual UVC prompted $\approx 50\%$ of the thrips to migrate toward other locations of the crop (Fig. 3B [in the original article] UVA + UVB and UVAr + UVB + UVCr). These results strongly suggest that thrips can directly detect and react behaviorally to natural and augmented UVB (Fig. 3B [in the original article]), even in a background of very strong visible and UVA radiation. Complementary experiments using UVB and UVA lamps (λ_{max} = 340 nm) again demonstrated a clear avoidance behavior induced by UVB (Fig. 3C [in the original article] UVAr + UVB), and suggested that thrips, like several other insects (Fig. 3C [in the original article] UVA), are somewhat attracted by UVA (See refs. 17, 21, and 28). In other words, the behavioral response induced by UVA was exactly the opposite of the response induced by UVB, demonstrating that perception and avoidance of UVB by thrips is not cued by UVA (see ref. 5). Our experiments cannot rule out the possibility that at least part of the behavioral effects of UVB in our system are mediated by the activation of photosensitizer compounds in the leaves coupled with the production of oxygen free radicals at leaf surface, which could be detected by the insects when they approach the boundary layer (see, e.g., ref. 29). However, if such a mechanism takes place in soybean leaves, the UV-absorbing compounds involved should have a fairly unusual specificity for UVB to explain the contrasting effects of UVB and UVA wavelengths on the insects (Fig. 3*C* [in the original article]). Regardless of the sensory mechanism, the observations of deleterious effects of solar UV exposure in other animal species (12, 13), suggest that the UVB-avoidance behavior that we have documented for thrips is likely to have fitness implications for the insects.

Insect-Mediated Effects on the Behavior of Other Insects. Dense thrips infestations can inflict serious damage to plants but, in our experiments, the effects of the thrips appeared to influence other consumers as well, like the soybean worm *A. gemmatalis*. Field and controlled-environment experiments in our laboratory have demonstrated that *Anticarsia* caterpillars, like thrips, consistently prefer soybean leaves grown under −UVB filters over leaves grown under full sunlight (unpublished results). Interestingly, however, in controlled-environment preference experiments, we found that caterpillars consistently avoided soybean leaves with even slight symptoms of previous thrips attack (Fig. 4 [in the original article]). Therefore, thrips damage, which is promoted by low UVB, also appears to induce antiherbivore responses in soybeans, an effect that may have parallels with the defense responses induced by insect attack and mechanical damage in other systems (see literature in refs. 27 and

30; see also refs. 31–34). It is important to point out, however, that because our experiments (Fig. 4 [in the original article]) made use of leaves that were naturally damaged by thrips, the possibility exists that the caterpillar preferences were determined by leaf traits that were inversely correlated with the severity of the thrips attack rather than by a product of the attack itself.

Conclusions

The magnitude of the effects of solar UVB on insect density in our field studies was very significant (Fig. 1 [in the original article]), especially considering that in some of the soybean crops used for these experiments we failed to detect any effects of solar UVB on crop gross morphology, UV-absorbing pigments, or biomass accumulation. Therefore, without negating the prevailing idea that plant responses to solar UVB are important for predicting ecosystem responses to a rise in UVB irradiances, our studies suggest that the possibility of direct effects of UVB on consumer behavior (Fig. 3 [in the original article]) deserves close examination. They also demonstrate that even in conditions in which plant growth responses to UVB are not detected, other as-yet-unidentified responses do take place and are sensed by phytophagous insects as part of their landing and host-selection clues (Fig. 2 [in the original article]). Variations in the behavior of grazing insects can have important effects on the dynamics of terrestrial food webs (ref. 35, and references therein). The effects of UVB on insect behavior that we have documented are likely to feed back on the primary producers and, more intriguingly, feed forward to other species of consumers (Fig. 4 [in the original article]). Exploring these networks of interactions in other systems will provide important clues on the ecological roles of solar UVB.

1. World Meteorological Organization. (1995) in *Report No. 37: Scientific Assessment of Ozone Depletion* (W.M.O., Geneva).

2. Caldwell, M. M., Teramura, A. H., Tevini, M., Bornman, J. F., Björn, L. O., & Kulandaivelu, G. (1995). *Ambio* 24, 166173 .

3. Rozema, J., Van de Staaij, J., Björn, L. O., & Caldwell, M. (1997). *Trends Ecol. Evol.* 12, 22–28.

4. Häder, D.-P., Worrest, R. C., Kumar, H. D., & Smith, R. C. (1995). *Ambio* 24, 174–180.

5. Bothwell, M. L., Sherbot, D. M. J., & Pollock, C. M. (1994). *Science* 265, 97–100.

6. Scientific Committee on Problems of the Environment. (1993) in *Effects of Increased Ultraviolet Radiation on Biological Systems* (SCOPE Secretariat, Paris).

7. Paul, N. D., Sharima Rasanayagam, M., Moody, S. A., Hatcher, P. E., & Ayres, P. G. (1997). *Plant Ecol.* 128, 296–308.

8. Koch, G. W. & Mooney, H. A., eds. (1996) in *Carbon Dioxide and Terrestrial Ecosystems* (Academic, San Diego).

9. Hatcher, P. E. & Paul, N. D. (1994). *Entomol. Exp. Appl.* 71, 227–233.

10. McCloud, E. S. & Berenbaum, M. R. (1994). *J. Chem. Ecol.* 20, 525–539.

11. Grant-Petersson, J. & Renwick, J. A. A. (1996). *Environ. Entomol.* 25, 135–142.

12. Blaustein, A. R., Hoffman, P. D., Hokit, D. G., Kiesecher, J. M., Walls, S. C., & Hays, J. B. (1994). *Proc. Natl. Acad. Sci. USA* 91, 1791–1795.

13. Williamson, C. E., Zagarese, H. E., Schulze, P. C., Hargreaves, B. R., & Seva, J. (1994). *J. Plankton Res.* 16, 205–218.

14. Malloy, K. D., Holman, M. A., Mitchell, D., & Detrich, H. W., III (1997). *Proc. Natl. Acad. Sci. USA* 94, 1258–1263.

15. Ballaré , C. L., Scopel, A. L., Stapleton, A. L., & Yanovsky, M. J. (1996). *Plant Physiol.* 112, 161–170.

16. Barcelo, J. A. & Calkins, J. (1979). *Photochem. Photobiol.* 29, 75–83.

17. Roberts, A. E., Syms, P. R., & Goodman, L. J. (1992). *Entomol. Exp. Appl.* 64, 259–268.

18. Smith, K. C. & Macagno, E. R. (1990). *J. Comp. Physiol.* A. 166, 597–606.

19. Tovée, M. J. (1995). *Trends Ecol. Evol.* 10, 455–460.

20. Greene, G. L., Leppla, N. C., & Dickerson, W. A. (1976). *J. Econ. Entomol.* 69, 487–488.

21. Stark, W. S. & Tan, K. E. W. P. (1982). *Photochem. Photobiol.* 36, 371–380.

22. Browman, H. I. & Hawryshyn, C. W. (1994). *J. Exp. Biol.* 193, 191–207.

23. Goldsmith, T. H. (1980). *Science* 207, 786–788.

24. Bennet, A. T. D., Cuthill, I. C., Partridge, J. C., & Lunau, K. (1997). *Proc. Natl. Acad. Sci. USA* 94, 8618–8621.

25. Jacobs, G. H., Neitz, J., & Deegan, J. F., II. (1991). *Nature* (London) 353, 655–656.

26. Zangerl, A. R., Arntz, A. M., & Berenbaum, M. R. (1997). *Oecologia* 109, 433–441.

27. Karban, R. & Baldwin, I. T. (1997) in *Induced Responses to Herbivory* (Univ. of Chicago Press, Chicago).

28. Antignus, Y., Mor, N., Joseph, R. B., Lapidot, M., & Cohen, S. (1996). *Pest Manage. Sampling* 25, 919–924.

29. Berenbaum, M. R. & Larson, R. A. (1988). *Experientia* 44, 1030–1031.

30. Bernays, E. A. & Chapman, R. F. (1994) in *Host Plant Selection by Phytophagous Insects* (Chapman & Hall, New York).

31. Barker, A. M., Wratten, S. D., & Edwards, P. J. (1995). *Oecologia* 101, 251–257.

32. Wold, E. N. & Marquis, R. J. (1997). *Ecology* 78, 1356–1369.

33. McConn, M., Creelman, R. A., Bell, E., Mullet, J. E., & Browse, J. (1997). *Proc. Natl. Acad. Sci. USA* 94, 5473–5477.

34. Agrawal, A. A. (1998). *Science* 279, 1201–1202.

35. Beckerman, A. P., Uriarte, M., & Schmitz, O. J. (1997). *Proc. Natl. Acad. Sci. USA* 94, 10735–10738.

6 The Sun as a Compass

Birds, like humans, possess an internal circadian rhythm clock that affects behavior throughout the day. A number of factors can influence this clock, but sunlight is the main driving force. In fact, some studies show that birds and humans left in complete darkness for days on end can lose their sense of time and become disoriented.

How is it that birds can migrate to specific parts of the world on a regular basis without a map or compass? Birds appear to be more sensitive to light changes and can use the Sun as their compass. This works well except at or near the polar regions, where longitude lines are closer together. The poles lead to faster time shifts, and those shifts can throw birds slightly off course. In the paper "Migration Along Orthodromic Sun Compass Routes by Arctic Birds," the authors track birds that migrate to and from high Arctic Canada. As predicted, the birds must adjust to quick internal clock changes. The findings support previous observations of birds that migrate between Siberia and the

Beaufort Sea. To do so, these birds must track the direction of the Sun's light as they cross the inhospitable Arctic Ocean. —JV

"Migration Along Orthodromic Sun Compass Routes by Arctic Birds"
by Thomas Alerstam, Gudmundur A. Gudmundsson, Martin Green, and Anders Hedenström
Science, January 2001

How birds use different compass systems based on the sun, stars, and geomagnetic field to orient along their migration routes is not fully understood (1–4). The region at the Northwest Passage where magnetic declination varies markedly close to the Magnetic North Pole provides a natural cue-conflict situation, in the sense that predicted flight trajectories differ in a very distinct way if birds orient by time-independent celestial rotation cues (predicted trajectories lie along geographic loxodromes) (2, 3, 5), the magnetic compass (predicted trajectories lie along magnetic loxodromes) (6), or the sun compass (7). The sun compass is sensitive to the time shift associated with longitudinal displacement (8–10), as long as the internal clock is not reset to local time, which apparently requires some period of stopover (11, 12). Such a natural time shift becomes substantial during migratory flights at polar latitudes, where distances between longitudes are small. At high latitudes, sun compass courses are expected to change by approximately 1° for each degree of longitudinal displacement, with little daily and seasonal variation. This corresponds to the change in sun azimuth associated with the difference in local time between longitudes (8). The course change along a great circle equals sin θ degrees for each degree of longitude intersected, where θ is the latitude (8). This means that at high latitudes, where sin θ approaches unity, sun compass routes constitute close approximations of orthodromes.

We have previously demonstrated regular and widespread east-northeast migration of shorebirds from northern Siberia toward North America across the Arctic Ocean (13). Evaluating predicted trajectories associated with different orientation principles showed orientation along sun compass routes to be most likely, although orientation along magnetic loxodromes could not be ruled out (14). In this study, we test this hypothesis of orientation along sun compass routes by investigating and evaluating the bird migration pattern in high arctic Canada, where there are extreme differences between predicted trajectories based on celestial and geomagnetic cues.

We used a tracking radar placed on board the Canadian Coast Guard icebreaker *Louis S. St-Laurent* to record the postbreeding bird migration pattern along the Northwest Passage during the period from 29 June to 3 September 1999 (15, 16). Radar observations were carried out when the ship was stationary (17), and more than 10 tracks of migrating birds were recorded at each of 11 sites between Baffin Island (66°W) in the east to Herschel Island (140°W) in the west. We focused on the dominating, high-altitude (mostly 400 to 3000 m above sea level), and broadfront migration in easterly and southeasterly directions as recorded at seven of these sites (18). Field observations from the ship and at tundra sites indicated that shorebirds (such as the American golden plover *Pluvialis dominica*, semipalmated sandpiper *Calidris pusilla*, white-rumped sandpiper *C. fuscicollis*, and pectoral sandpiper *C. melanotos*) made up most of the migrants in this study, traveling in flocks up to about 100 individuals. The above-mentioned species of shorebirds have a wide breeding range in arctic North America, migrating mainly via the Atlantic region of North America to well-defined winter quarters in South America. Hence, the results from the different sites are expected to reflect, to a large extent, this major shorebird migration system in the New World (19).

The mean and scatter of flight directions are shown in Table 1 [in the original article]. Track direction refers to the flight direction relative to the ground, as measured by radar, whereas heading direction was calculated by subtracting the wind vector from the birds' track vector. We primarily refer to observed track directions in our evaluation. The differences between track and heading directions are generally small or modest (Table 1 [in the original article]), and our conclusions will be valid irrespective of whether partial wind drift has influenced the track directions on some occasions.

Eastbound migration was massive at all three Beaufort sites (combined in Table 1 [in the original article]), with similar mean track directions 100 km north of the coast (105° at 70.5°N, 139.0°W), 27 km north of the coast (87° at 69.8°N, 133.3°W), and only a few kilometers from Herschel Island (104° at 69.6°N, 139.5°W). Combining all eastbound tracks (0° to 180°) from these three sites, 33, 42, and 16% fell in the sectors 60° to 90°, 90° to 120°, and 120° to 150°, respectively. Selected courses within these sectors have been used to illustrate predicted trajectories to and from the Beaufort sites for different orientation principles, whereas corresponding trajectories were calculated on the basis of mean directions for the Wollaston, King William, and Baffin sites (Fig. 1 [in the original article]). When evaluating the alternative trajectories, we rejected those that do not take the birds through the Hudson Bay region or the interior and east coast of North America, where these shorebirds are known to pass during autumn migration (19).

Long-distance orientation along constant geographic or magnetic courses can be ruled out, because the geographic and magnetic loxodromes (20) in most cases extend too far north, with some of the latter spiralling toward the Magnetic North Pole (Fig. 1 A through D, [in the original article]). Geographic loxodromes from the Beaufort sites toward 60 to 110° (65% of the recorded tracks), as well as the loxodrome along the mean direction at Wollaston Peninsula, can be rejected (Fig. 1, A and B, [in the original article]). All magnetic loxodromes except one (Baffin Island) can also be rejected. Magnetoclinic orientation (21) is also invalid in many cases (14). It has been recently suggested, on the basis of experiments testing the interactions between celestial and magnetic compasses in birds, that the problem of a changing magnetic declination is solved by recalibration of magnetic orientation by celestial rotation cues at stopover sites along the route (2). This is probably not a feasible solution for the shorebirds in this study, because it necessitates very frequent recalibration (eastbound movement between 120° and 90°W at the 70°N parallel is perpendicular to densely spaced declination isolines involving a change in magnetic declination by about 70° per 1000 km of distance, and inclination is very steep, 85° to 88°) and still leads the birds along unrealistic loxodromes in many cases.

Sun compass trajectories (Fig. 1, E and F [in the original article]) show an excellent agreement with known autumn concentration of shorebirds at, for example, James Bay and the east coast of North America, and with the transatlantic migration to South America (19, 22–25). These trajectories also agree with a migration system between Siberia and North America across the Arctic Ocean (14). Only for the migration at Baffin Island do all three orientation principles give similar trajectories, and no conclusion about the most likely alternative can be drawn for this site (Fig. 1, B, D, and F [in the original article]). The fact that sun compass routes are distance-efficient is illustrated in Fig. 2 [in the original article], showing trajectories on an equidistant azimuthal map projection (26). For this evaluation, it is the predicted trajectories at northerly latitudes that are critical. The birds may change to other orientation principles and cues for the transoceanic flights across the Atlantic Ocean (23–25) and at more southerly latitudes, where celestial and geomagnetic cues (and wind patterns) are quite different from those at high latitudes

[Sun compass trajectories in Figs. 1 and 2 (in the original article) are extended across the Atlantic Ocean and into more southerly latitudes (8) without implying any critical evaluation in relation to alternative possibilities of orientation in these latter regions].

Orientation along sun compass routes is facilitated by long nonstop flights (common among arctic shorebirds), when there is no time for the birds to reset their internal clock in phase with local time. To continue along the same sun compass route after a stopover period when the birds reset their internal clock to local time requires that the arrival course at the stopover locality be transferred from the sun compass to another cue system (such as magnetic compass or topography), to be transferred back to the sun compass as the departure course after resetting of the internal clock (8).

Although the radar tracks showed that migration takes place on a broad front, apparently without responses to landmark features, large-scale topographical guidance could play some role in the birds' flight routes (such as along the south coast of the Beaufort Sea). However, there are long passages, across ocean, barrens, or low-level fog, where the migrants cannot rely on topographical orientation. Our analysis does not exclude the possibility that the birds use a more complex orientation program than assumed in the alternative cases above, perhaps involving changes of preferred compass course and/or switching between different celestial and magnetic orientation mechanisms during the passage across arctic North America. Genetically programmed changes in migratory direction have been demonstrated for passerines (1, 5) and it is also known that migratory birds may switch between magnetic and celestial orientation cues (1–3). Even if feasible in principle, it is difficult to propose a complex orientation program that accounts for the observed migration patterns under the special geomagnetic and environmental conditions at the Northwest Passage. Furthermore, the orientation from Siberia toward North America across the Arctic Ocean (14) would require the assumption of an analogous complex system programmed in relation to quite different geomagnetic and environmental conditions, which seems highly unlikely. Also, the long flights and rapid postbreeding migration from the tundra to temperate staging areas, which are common features among shorebirds (the passage across the Arctic Ocean from Siberia to North America is presumably completed in a single flight), make a division of the journey into a series of differently programmed orientation steps unlikely. Before considering such intricate solutions, we investigated whether fixed orientation on the basis of any of the birds' fundamental compass mechanisms may suffice to explain the long-distance orientation and migration routes at high magnetic and geographic latitudes.

This analysis supports the hypothesis of sun compass routes as an explanation for the long-distance orientation of many tundra shorebirds between Siberia, arctic North America, and the east coast of North America, whence the birds continue by transatlantic flights toward South American winter quarters. An important feature of this migration system is the fact that the resulting sun compass routes conform closely to distance-saving orthodromes.

The sun compass has probably provided the basis for the evolution of the above-mentioned extraordinary migration system in a region with an exceptional geomagnetic field. Along routes where magnetic declination is less variable, birds may use their interacting compass senses differently (27, 28). The sun compass has been demonstrated to be important to Adelie penguins (*Pygoscelis adeliae*) in Antarctica (9, 10)—another case of polar orientation. The fact that the arctic shorebirds in this study do not return in spring along the same routes as used in autumn [spring migration mostly occurs through the interior of North America (19)] testifies to the complexity of the global orientation performance of migrating birds.

References and Notes

1. P. Berthold, Ed., *Orientation in Birds* (Birkhäuser Verlag, Basel, Switzerland, 1991).

2. K. P. Able, M. A. Able, *Nature* 375, 230 (1995).

3. W. Wiltschko, P. Weindler, R. Wiltschko, *J. Avian Biol.* 29, 606 (1998).

4. R. Wehner, *J. Avian Biol.* 29, 370 (1998).

5. S. T. Emlen in *Avian Biology, Volume V*, D. S. Farner, J. R. King, Eds. (Academic Press, New York, 1975), pp. 129–219.

6. R. Wiltschko, W. Wiltschko, *Magnetic Orientation in Animals* (Springer-Verlag, Berlin, 1995).

7. G. Kramer, *Ibis* 99, 196 (1957).

8. T. Alerstam, S.-G. Pettersson, *J. Theor. Biol.* 152, 191 (1991).

9. J. T. Emlen, R. L. Penney, *Ibis* 106, 417 (1964).

10. R. L. Penney, J. T. Emlen, *Ibis* 109, 99 (1967).

11. K. Hoffmann, *Z. Tierpsychol.* 11, 453 (1954).

12. K. Schmidt-Koenig, *Z. Tierpsychol.* 15, 301 (1958).

13. T. Alerstam, G. A. Gudmundsson, *Arctic* 52, 346 (1999).

14. T. Alerstam, G. A. Gudmundsson, *Proc. R. Soc. London Ser. B* 266, 2499 (1999).

15. T. Alerstam, G. A. Gudmundsson, M. Green, A. Hedenström, B. Larsson, in *Cruise Report Tundra Northwest 1999*, E. Grönlund, Ed. (Swedish Polar Research Secretariat, Stockholm, Sweden, 2000), pp. 122–126.

16. The icebreaker route extended from Iqaluit on Baffin Island (29 June 1999) through Hudson Strait, Foxe Basin, and northward to northern Bathurst Island, where the ship turned south- and westward, passing south of Victoria Island to Herschel Island at the Beaufort Sea as the westernmost position (3 to 5 August). The ship then sailed eastward by another route north of Banks Island via the Magnetic North Pole at Ellef Ringnes Island (79°N, 105°W), returning to Iqaluit (3 September) along the east coast of Baffin Island.

17. A tracking radar (3 cm wavelength, 200 kW peak power, 0.25/1.0 μs pulse duration, and 1.5° nominal pencil beam width) was placed on board the CCG icebreaker *Louis S. St-Laurent* (radar antenna 21 meters above sea level). The radar was operated when the ship was stationary in pack ice or open water, most often 2 to 8 km from the nearest tundra shores, except at two sites in the Beaufort Sea, 27 and 100 km north of the coast, respectively. The position of a target was recorded every 2 s by the radar operated in automatic tracking mode. Radar data were corrected for the heading, motion, and leveling of the ship recorded simultaneously with the radar tracking. Mean track direction, ground speed, vertical speed, and altitude were calculated for flocks or individual birds tracked for at least 30 s (mean tracking time was 160 s) within 10 to 12 km from the radar. Wind measurements at different altitudes were obtained within 2 hours of bird track records by radar tracking of helium-filled balloons. Mean heading direction and airspeed of the birds were calculated by subtraction of the wind vector from the birds' track vector. For some tracks, heading directions could not be calculated because of missing wind data.

18. A grand total of 692 radar tracks of birds was recorded during 494 hours of radar operation during the journey. In this analysis, we focused on a well-defined and intensive east- and southbound migratory passage recorded at seven sites, excluding sites with sparse (≤10 radar tracks) or widely scattered movements (three sites at Banks and Melville Islands) or mainly low altitude (< 200 m) movements over the sea (Amundsen Gulf). Included in our analyses are three sites at the Beaufort Sea with massive eastbound migration (279 tracks toward 0° to 180°; 23 tracks toward 180° to 360° excluded), one site at Wollaston Peninsula (21 tracks toward 0° to 180°; 4 tracks toward 180° to 360° excluded), one site at King William Island (18 tracks toward 0° to 180°; at this site there was a distinctly bimodal distribution toward SE and W, and 32 tracks toward 210° to 350° were excluded), and two sites at Baffin Island (50 tracks towards 90° to 270°; 4 tracks toward 270° to 90° excluded).

19. R. I. G. Morrison, in *Behavior of Marine Animals, Vol 6. Shorebirds: Migration and Foraging Behavior*, J. Burger, B. L. Olla, Eds. (Plenum, New York, 1984), pp. 125–202.

20. J. M. Quinn, GEOMAG (U.S. Geological Survey, Denver, CO, 2000). Program software available at http://geomag.usgs.gov/frames/geomagix.htm. Magnetic loxodromes and declination chart were calculated on the basis of geomagnetic data for July–August 1999 according to the models WMM-95 and WMM-2000.

21. J. Kiepenheuer, *Behav. Ecol. Sociobiol.* 14, 81 (1984).

22. R. McNeil, J. Burton, *Wilson Bull.* 89, 167 (1977).

23. T. C. Williams, J. M. Williams, in *Animal Migration, Navigation, and Homing*, K. Schmidt-Koenig, W. T. Keeton, Eds. (Springer-Verlag, Berlin, 1978), pp. 239–251.

24. W. J. Richardson, *Can. J. Zool.* 57, 107 (1979).

25. P. K. Stoddard, J. E. Marsden, T. C. Williams, *Anim. Behav.* 31, 173 (1983).

26. G. A. Gudmundsson, T. Alerstam, *J. Avian Biol.* 29, 597 (1998).

27. G. A. Gudmundsson, R. Sandberg, *J. Exp. Biol.* 203, 3137 (2000).

28. R. Sandberg, J. Bäckman, U. Ottosson, *J. Exp. Biol.* 201, 1859 (1998).

Bees, like birds, are highly attuned to the Sun's course. They use the Sun as a compass to travel over tremendous distances for food. Honeybees also use directional information provided by the Sun to communicate food locations to other members of their hive. Honeybees are known for their superior eyesight, but is seeing believing when it comes to the Sun?

In "Development of Sun Compensation by Honeybees: How Partially Experienced Bees Estimate the Sun's Course," the authors investigate whether Sun-sight deprived bees could still engage in accurate navigation and precise communication of their navigation by hive waggle dances. As you will learn, no matter how much the scientists limited the test honeybees' view of the Sun, the bees still moved and danced with near-perfect knowledge of the Sun's course. The findings suggest that honeybees have an innate knowledge of solar movement that enables them to estimate the Sun's course even when they do not observe it. While reading the paper, also keep in mind that honeybees are some of the most intelligent creatures in the insect world. It is not easy to fool a bee! —JV

"Development of Sun Compensation by Honeybees: How Partially Experienced Bees Estimate the Sun's Course"
by Fred C. Dyer and Jeffrey A. Dickinson
Proceedings of the National Academy of Sciences,
May 1994

A wide variety of animals can use the sun's azimuth as a true compass, compensating for its daily movement relative to terrestrial features (1–3). Complicating this task is that the azimuth changes at a variable rate over the day, shifting relatively slowly as the sun rises in the morning or sets in the evening and rapidly as the sun crosses the local meridian at midday. Furthermore, the daily pattern of change, the ephemeris function, varies with season and with latitude. Animals are generally thought to learn the current local ephemeris function, rather than using the average rate of shift of the azimuth. What little is

known about this learning process suggests that animals do not merely memorize a series of time-linked solar positions. Instead, in a wide range of invertebrate (4–12) and vertebrate (13–15) taxa, animals have been shown to estimate solar positions that they have never seen. For example, in a classic series of experiments, Lindauer (4, 5) showed that incubator-reared honeybees (*Apis mellifera*) that were allowed to see only the western half of the sun's course (from noon onward each day) could, when they flew for the first time in the morning, use the sun to search for food in a direction in which they had been trained. Thus, somehow they had correctly learned to expect the sun in the eastern half of the sky in the morning. In this paper we ask what integrative mechanism might underlie this apparent ability of bees to compute solar positions that fill gaps in their experience of the sun's course.

Four distinct computational strategies have been proposed to account for the ability of insects to estimate unknown positions of the sun. Three of these assume the existence of neural computations that operate on time-linked measurements of the sun's position relative to the landscape, to calculate a compensation rate specific to each gap in the animal's experience that needs to be filled. First, the animal might "interpolate" at a linear rate to estimate the sun's position between temporally adjacent known positions. Second, she might "extrapolate" the sun's course forward at a linear rate based upon the azimuth and rate of movement that she has observed most recently. Third, she might extrapolate "backward" at a linear rate based upon the azimuth and rate of movement that she has observed at a later hour on previous days. Each of these hypothesized mechanisms could be used to assemble, through repeated measurements of the sun's position over the day, a representation of the current local ephemeris function. Although studies of honeybees (*Apis spp.*) (6, 16) and desert ants (*Cataglyphis spp.*) (8, 17) have tended to favor the linear interpolation hypothesis, other data seem best accounted for by the forward (9, 18) and backward (9) extrapolation hypotheses. Some data, including Lindauer's (4, 5) observations, are consistent with all of these proposed mechanisms.

A more recent study of *Cataglyphis fortis* (19) suggested an entirely new hypothesis: that the ants might rely upon an innate knowledge of the general pattern of solar movement, which allows individuals with limited experience to develop an approximate representation of the sun's entire course. The approximate representation is assumed to be adjusted by experience to match the sun's actual course more closely. Experimental data decisively excluded the linear interpolation and extrapolation hypotheses for this ant; the evidence for the new hypothesis, however, was open to other interpretations (see ref. 19).

We developed an approach to this problem that is inspired by Lindauer's studies of experience-restricted bees but that relies upon observations of dance orientation rather than flight orientation. The bees' communicative waggle dances encode in their orientation the direction of food relative to the sun and thus reflect the bees' determination of the current solar azimuth (20). By testing bees on cloudy days, we forced them to rely upon an internal estimate of the solar azimuth instead of a direct measurement. Fully experienced bees orient their dances on cloudy days by drawing upon an accurate memory of the sun's entire course relative to familiar features of the terrain (16, 21). We hoped that dances by experience-restricted bees would reveal how they estimated the sun's position when they flew under cloudy skies at a new time of day.

Methods

We performed experiments with two small colonies formed entirely of bees that had emerged from brood comb in an incubator. The experimental bees were allowed out of the hive only during the last ≈4 hr of daylight each afternoon, so that they could see the sun over only 45–50° (≈20%) of its diurnal course. During the daily flight period bees were trained to visit a feeder offering sugar syrup 350 m south of the hive (along a line of tall trees to provide a reliable reference for flight orientation and for learning the sun's course; refs. 16 and 21) and were given identifying labels. During times other than the flight period, colony 1 was moved indoors, where diffuse light was provided through a window to expose bees to the natural photoperiod. Colony 2 remained in place at all times and was set up to provide differing experience to two groups of bees. One group, which served as experience-restricted bees, was labeled on the thorax with large plastic tags, while a second group, which was allowed to fly throughout the day, was labeled with dots of paint. Modifying a method developed by G. S. Withers, S. E. Fahrbach, and G. E. Robinson (personal communication), we placed a metal grating over the hive entrance to prevent the departure of tagged bees, but not paint-dotted or untreated bees, until the beginning of the daily flight period. In both colonies, the experience-restricted foragers flew during the afternoon training period for at least 7 days prior to the test; Lindauer (5) found that bees with afternoon experience can estimate the morning position of the sun after 3–5 days of experience. On the day of a test, we allowed all bees out of the hive in the morning under a cloudy sky and through a window in the hive observed their dances after they returned from the training feeder. The orientations of waggling runs relative to gravity were measured with a protractor.

The cloud cover during both experiments was thick and homogeneous and produced almost continuous light rain. Clouds depolarize sky light and block the sun from view, thus depriving bees of celestial orientation cues. Bees were once thought to determine their orientation relative to the sun on cloudy days primarily by detecting the sun directly in the ultraviolet (22). More recent studies, however, suggest that bees should rarely, if ever, be able to see the sun when it is invisible to human observers. Physical measurements of radiance patterns across an overcast sky (23) failed to detect the sun in any wavelength, unless it became visible to the human eye. Also, observations of bees dancing on overcast days showed that the sun or blue sky is virtually always visible to human observers when bees base their dances on direct measurements of celestial cues during the flight rather than on memory (16); we could detect no such inhomogeneities in the cloud cover during our experiments.

Results

The dance orientations of the experience-restricted bees, which had never before flown or danced during the morning or early afternoon, fell into a striking pattern that crudely approximated the actual course of the sun (Figs. 1 and 2 [in the original article]). With considerable consistency among bees, and among different dances by the same bees, dances performed in the morning suggested that bees expected the sun to be almost exactly 1800 from its position in the late afternoon. The bees maintained this orientation until a short interval at midday when, as a group, they changed their dance angles by 180°. (The shift occurred about an hour earlier in colony 1 than in colony 2, perhaps because bees in the two colonies experienced slightly different photoperiod cues.) After the shift the bees maintained the new orientation with little change. This same pattern was exhibited by bees that performed many dances throughout the day and thus does not represent two populations of differently oriented bees. Strikingly, the only exceptions to the pattern were two bees in colony 1 that adopted the evening angle in the morning and the morning angle in the afternoon (Fig. 1 [in the original article]).

The linear interpolation, forward extrapolation, and backward extrapolation hypotheses failed to predict the particular orientations adopted by dancers at most times of day, and none of these hypotheses predicted the abrupt shift in orientation at midday. Conceivably some combination of these hypotheses explains the bees' behavior, but we believe it more parsimonious to reject all three, at least for the conditions of this experiment.

Another possible interpretation of the data, given that the solar positions estimated by dancers usually did not deviate enormously from the actual azimuth, is that bees obtained during the flight a skewed, but nevertheless direct, measure of the sun's position. As discussed, the completeness of the cloud cover during our experiment should have precluded this possibility. Also, it seems most unlikely that bees extracting a weak visual signal from the noise of the overcast sky should have been so consistent in their measurement of the sun's position. Finally, a control is provided by our experiment with colony 2 (Fig. 2 [in the original article]), in which experience restricted bees conformed to the basic pattern already described, while bees that had previously foraged throughout the day tracked the course of the sun more accurately, as expected from previous work (16). Had celestial cues been visible through the cloud cover to bees on their flights preceding the dances, the behavior of the two groups, which flew together under the same sky, might have been similar. Thus, we conclude that both groups internally generated an estimate of the sun's position relative to local landmarks, but differed because of the extent of their past experience.

The bee's behavior over the day is well described by a simple step function in which a single morning angle is replaced by a single afternoon angle at midday (10:50 for colony 1, 12:00 for colony 2), such that the morning angle is exactly 180° from the azimuth experienced at the middle of the afternoon flight period (270° for colony 1, 256° for colony 2) (Fig. 3 [in the original article]). For the experience-restricted bees (excluding the two exceptional bees in colony 1, which seemed to use a 180° step function that was out of phase by 12 hr or 1800) the step function accounts for the variance in the data significantly better than does the actual ephemeris function (colony 1: F_s = 2.011, df = 536, 536, P < 0.001; colony 2: F_s = 2.764, df = 179, 179; P < 0.001). For the fully experienced bees in colony 2, by contrast, the actual ephemeris function provides a significantly better fit (F_s = 2.637, df = 59, 59, P < 0.001).

Discussion

Our results decisively eliminate three previously proposed computational hypotheses to explain how honeybees estimate unknown portions of the sun's azimuthal course. We suggest instead that bees are innately informed of certain general spatial and temporal features of solar movement. This allows partially experienced bees to construct an internal representation that not only reflects the portions of the sun's course previously seen but also approximates the dynamics of solar movement at other times of day. In the bees' approximation, as in the actual ephemeris function, (*i*) the sun's position at dawn is about 180° from its position near sunset (as measured relative to local landmark features), (*ii*) the rate of change of the azimuth is similar in the morning and in the afternoon, and (*iii*) the sun moves from the eastern to the western half of the sky at midday. With experience restricted to the late afternoon, the bees' representation resembles a 180° step function. With further experience it conforms more closely to the actual ephemeris function.

Strikingly similar behavior has recently been reported for desert ants (*Cataglyphis fortis*), observed while using the sun for homing (19). Ants that had previously left the nest only in the morning behaved when tested in the afternoon as if they expected the sun to be about 180° from its morning position. However, the testing was mainly restricted to the late afternoon and so did not reveal how ants represented the sun's course throughout the day. Also, the data could not decisively exclude the faint possibility that the experience-restricted ants had previously observed the sun's afternoon azimuth from the nest entrance (19). Our experiments not only strongly indicate that honeybees with restricted prior experience can derive an approximate representation of the sun's overall pattern of movement but also provide a clear picture of the accuracy and dynamics of this representation.

This mechanism of sun-azimuth learning suggested by the behavior of experience-restricted bees and ants would offer certain advantages for a short-lived, small-brained animal. Most important, in comparison with the interpolation and extrapolation models outlined earlier, it would allow individuals to acquire a relatively accurate representation of the solar ephemeris function, without needing to sample more than a small portion of the sun's course. In some circumstances the accuracy could be scarcely improved by additional experience. For example, during some seasons at low latitudes, the solar ephemeris function closely resembles the step function used by afternoon-experienced bees and (presumably) by morning-experienced ants: the azimuth changes little during the morning, then switches abruptly by about 180° as the sun crosses the meridian, then changes little during the afternoon (Fig. 3 [in the original article]).

A step function gives a good description of our data, but the actual operations that underlie the development of a bee's internal ephemeris remain to be determined. In fact, evidence against the notion that bees represented the sun's course literally as a step function was provided by the behavior of the few bees in colony 1 ($n = 4$) that danced repeatedly during the rapid transition at midday. All changed their orientations progressively rather than instantaneously, as if they had represented the sun's course as a continuous function. Two bees behaved as if the sun shifted clockwise across the southern horizon and two behaved as if it shifted counterclockwise across the northern horizon, implying that they had obtained conflicting or incomplete information during the training period about which direction the sun was "supposed" to move. These observations reinforce the conclusion that bees can estimate global properties of solar movement that they have never seen, as if they have an innate 'template" (24) guiding the learning process. They also underscore the importance of experience in fine-tuning the spatial and temporal correspondence between the bees' representation of the sun's course and the actual ephemeris function.

The hypothesis that bees (and ants; ref. 19) are innately informed of the general pattern of solar movement may account for other results that have heretofore been interpreted in other ways (6, 8, 9, 16, 17). Most strikingly, it may explain observations of bees dancing in the tropics on days when the sun's arc passed within 5° of the zenith at local noon (6). Apparently unable to resolve an azimuth during the ≈0.5 hr spanning local noon, bees behaved as if they estimated the sun's midday course. They rotated successive dances progressively but rapidly to compensate for the ≈180° change of the azimuth; some bees turned steadily clockwise, some counterclockwise. This finding had suggested that bees compensated by linear interpolation between the solar positions in the morning (east) and afternoon (west) and that their direction of compensation was ambiguous because the gap to be filled was 180°. That our bees exhibited similar behavior without having seen the morning or midday sun suggests that linear interpolation need not be invoked for the midday behavior of bees in the tropics. It remains to be determined whether the interpolation and extrapolation mechanisms outlined earlier play any role in learning the current ephemeris function. Whatever computational processes are involved, the method we have outlined offers a promising route to exploring them in more detail.

1. Wehner, R. (1984) *Annu. Rev. Entomol.* 29, 277–298.

2. Able, K. P. (1980) in *Animal Migration*, ed. Gauthreaux, S. A. (Academic, New York), pp. 283–373.

3. Schmidt-Koenig, K., Ganzhorn, J. U. & Ranvaud, R. (1991) in *Orientation in Birds*, ed. Berthold, P. (Birkhiuser, Basel), pp. 1–15.

4. Lindauer, M. (1957) *Naturwissenschaften* 44, 1–6.

5. Lindauer, M. (1959) *Z. Vgl. Physiol.* 42, 43–62.

6. New, D. A. T. & New, J. K. (1962) *J. Exp. Biol.* 39, 271–291.

7. Edrich, W. (1981) *Physiol. Entomol.* 6, 7–13.

8. Wehner, R. (1982) *Neujahrsbl. Naturforsch. Ges. Zurich* 184, 1–132.

9. Dyer, F. C. (1985) *Anim. Behav.* 33, 769–774.

10. Pardi, L. (1957/58) *Atti Accad. Sci. Torino* 92, 65–72.

11. Pardi, L. (1953-54) *Boll. Ist. Mus. Zool. Univ. Torino* 4, 127.

12. Birukow, G. (1956) *Z. Tierpsychol.* 13, 463–484.

13. Braemer, W. (1959) *Verh. Dtsch. Zool. Ges.* 19S9, 276–288.

14. Hoffmann, K. (1959) *Z. Vgl. Physiol.* 41, 471–480.

15. Schmidt-Koenig, K. (1961) *Naturwissenschaften* 48, 110.

16. Dyer, F. C. (1987) *J. Comp. Physiol. A* 160, 621–633.

17. Wehner, R. & Lanfranconi, B. (1981) *Nature (London)* 293, 731–733.

18. Gould, J. L. (1980) *Science* 207, 545-547.

19. Wehner, R. & Muller, M. (1993) *Naturwissenschafien* 80, 331–333.

20. Frisch, K. v. (1967) *The Dance Language and Orientation of Bees* (Harvard/Belknap, Cambridge, MA).

21. Dyer, F. C. & Gould, J. L. (1981) *Science* 214, 1041–1042.

22. Frisch, K. v., Lindauer, M. & Schmeidler, F. (1960) *Naturwissensch. Rundsch.*, 169–172.

23. Brines, M. L. & Gould, J. L. (1982) *J. Exp. Biol.* 96, 69–91.

24. Marler, P. (1984) in *The Biology of Learning,* eds. Marler, P. & Terrace, H. S. (Springer, Berlin), pp. 289–309.

Solar Energy

The Sun provides a seemingly infinite amount of free energy, yet most of this energy now is wasted. Every day, the Sun releases 15,000 times more energy than humans use within that same time frame. The rest of the energy either dissipates as heat or simply is not collected. While there is a lot of monetary investment in the use of fossil fuels, many researchers are investing their time toward the development of effective, practical solar energy systems and technologies.

In "Thin Films Seek a Solar Future: Despite Setbacks, the Technology May Yet Shine," Ineke Malsch describes a silicon product that appears to be cost-effective and relatively easy to mass produce. The invention, a crystalline thin film, consists of multilayered cells that capture sunlight's heat and energy and, with other equipment, converts this energy into electricity. As the article points out, such research can benefit humans because reliance upon fossil fuels has led to the emission of pollutants like carbon dioxide. The Sun, on the other hand, produces *energy without pollution. If we can harness that energy, we will have an unlimited source of power for the next 5 billion or so years. —JV*

"Thin Films Seek a Solar Future: Despite Setbacks, the Technology May Yet Shine"
by Ineke Malsch
The Industrial Physicist, April/May 2003

Energy from the sun—available everywhere, for every-body—has motivated research on solar-energy technologies for about three decades. The U.S. Photovoltaic Industry Roadmap, intended to guide companies in developing solar-energy systems, takes a more prosaic but realistic view of the next three decades. It aims for solar energy to provide 10 % of U.S. peak generation capacity and supply a considerable share of foreign markets by 2030.

Most photovoltaic (PV) solar technologies rely on semiconductor-grade crystalline-silicon wafers, which are expensive to produce compared with energy from fossil fuel sources. However, potentially less costly thin-film alternatives may make major inroads in the world market in five years, suggests Franz Karg, research manager at the Shell Solar facility in Munich, Germany. Or maybe not. Thin-film solar panels are hard to mass-produce cost-effectively because of the difficulty of coating large areas of glass. "It is my opinion that crystalline-silicon technologies will dominate for at least the next 10 years," says Jeffrey Mazer of the U.S. Department of Energy (DOE) Office of Solar Energy Technologies (Washington, DC).

Solar-energy systems pose many challenges for developers, particularly in the current world economy. Last October, Shell Solar (Amsterdam, The Netherlands) announced the closing of two production operations—in Helmond, The Netherlands, and in Munich—in a restructuring meant to make the company more competitive. The next month, BP Solar (Linthicum, MD) decided to close down production of its thin-film amorphous silicon and cadmium telluride (CdTe) solar panels to focus on crystalline-silicon technologies. "While the thin-film technology continues to show promise, lack of present economics does not allow for continued investment," said Harry Shimp, BP Solar's president and chief executive officer.

BP's decision is a setback to the marketing of new thin-film solar technologies. However, First Solar, LLC (Perrysburg, OH), a major maker of CdTe solar cells, remains strongly committed to the technology. Shell Solar continues development of its thin-film technologies. And DOE's National Renewable Energy Laboratory (NREL) in Golden, Colorado, continues to provide funds for thin-film PV research to its 40 industrial and university partners.

Why Solar Power?

In 2001, the global market for PV panels and equipment was valued at $2 billion. Worldwide in 2000, solar, geothermal, wind, combustible renewables, and burning garbage and other wastes collectively provided 1.6% of electricity production, according to the International Energy Agency (IEA) in Paris. In its World Energy

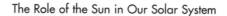

Outlook 2002, IEA described two scenarios for world energy demand and supply until 2030. One scenario assumes only the continuation of current government measures to stimulate sustainable-energy supply and demand. In it, fossil fuels continue to meet more than 90 % of energy demand. All renewables except hydropower grow by 3.3 % annually, but they will not meet a large share of the total energy demand because of their low-percentage base. Carbon dioxide (CO_2) emissions will grow 70 % by 2030 to 38 billion metric tons annually.

In the alternative scenario, governments will implement policies such as promoting energy efficiency, the use of cleaner energy sources, and reducing the environmental impact of producing and burning fossil fuels. Compared with 1990, those changes would result in an estimated 16 % fewer CO_2 emissions in 2030—a year when the United Nations estimates the world population will be about 8.3 billion—in part because renewable energy sources will grow rapidly.

Current applications of PV solar panels include providing power to spacecraft and isolated villages in developing countries, solar-energy systems in homes and buildings in Western countries (Figure 1 [in the original article]), and even powering the lamps of remote lighthouses. "Especially where there is no connection to the grid, solar energy is easily cheaper than small-scale electricity production with a diesel generator, to give an example," Karg says. "For electricity production in rural areas in developing countries, solar is the cheaper alternative. To achieve more, we need breakthroughs in large-scale storage of electricity, and solar must be developed in combination with wind, biomass, energy-storage systems, and fossil fuels."

Most people in solar energy consider government subsidies for R&D and sales as necessary for its successful development and increased usage. Indeed, the PV energy business is still largely dependent on government intervention, and most U.S., European, and Japanese projects are subsidized. The Bush administration seeks $79.7 million from Congress for fiscal year (FY) 2004 to support solar-energy research, up 0.1% from its amended FY 2003 request but down from the $87.1 million that Congress appropriated in FY 2002.

Crafting Photovoltaics

PV solar panels convert sunlight directly into electricity. A panel consists of several connected 0.6-V dc PV cells, which are made out of a semiconducting material sandwiched between two metallic electrodes. "The photovoltaic effect refers to the separation of minority carriers [electrons and holes] by a built-in electric field," such as a pn-junction or Schottky barrier, says DOE's Mazer. The cells are usually encapsulated behind glass to weatherproof them. In a PV array, several panels are connected to provide sufficient power for common electrical applications such as household electricity. The arrays can be connected to an electricity grid or work as standalone systems.

Researchers at what is now Lucent Technologies' Bell Laboratories first demonstrated silicon solar cells in 1954, and most PV systems today use mono- or multicrystalline

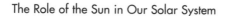

silicon as the semiconducting material. "We obtain monocrystalline wafers by sawing them from silicon rods, which we grow by the Czochralski growth process," explains Ronald van Zolingen, professor of sustainable energy at the Technical University in Eindhoven, The Netherlands, and a senior business advisor to Shell Solar. "In this process, we pull a monocrystalline rod from a liquid, starting with a small crystal. The growth speed is relatively low, but we obtain excellent material. Monocrystalline silicon solar cells have the advantage of a high efficiency, about 15 %, which is an advantage for specific applications."

"We obtain multicrystalline wafers from ingots grown by casting liquid silicon in a large container followed by controlled cooling," van Zolingen adds. "This technique is less complicated than the pulling of single-crystalline rods. Multicrystalline-silicon solar cells have a slightly lower efficiency than monocrystalline, about 13.5 %." Worldwide, the production of multicrystalline-silicon solar cells outpaces that of monocrystalline-silicon solar cells.

Among the major bottlenecks to the output of crystalline-silicon PVs is the high loss of materials during production of the wafers. "In addition, we need to saw the silicon. We typically lose 0.2 mm at the kerf," says van Zolingen. Despite these problems, crystalline silicon remains the dominant solarcell material. One reason for this is the support provided by the federal government's PV program since the 1970s for R&D projects focused on crystalline-silicon technologies.

Thin-film alternatives to standard PV solar cells are already available or in development. Amorphous silicon, the most advanced of the thin-film technologies, has been on the market for about 15 years. It is widely used in pocket calculators, but it also powers some private homes, buildings, and remote facilities. An amorphous-silicon solar cell contains only about 1/300th the amount of active material in a crystalline-silicon cell. Amorphous silicon is deposited on an inexpensive substrate such as glass, metal, or plastic, and the challenge is to raise the stable efficiency. The best-stabilized efficiencies achieved for amorphous-silicon solar panels in the U.S. PV program are about 8%. The goal is to produce a stable device with 10% efficiency. United Solar Systems Corp. (Troy, MI) pioneered amorphous-silicon solar cells and remains a major maker today.

Thin-film crystalline-silicon solar cells consist of layers about 10 µm thick compared with 200- to 300-µm layers for crystalline-silicon cells. Researchers at NREL use porous polycrystalline silicon on low-cost substrates and trap light in the silicon to enable total absorption. They have fabricated working solar cells with this material.

New Thin Films

Copper indium diselenide (CIS) is a more recent thin-film PV material. Siemens Solar developed a process for depositing layers of the three elements on a substrate in a vacuum, and Shell Solar later acquired the technology when it bought Siemens Solar (Figures 2 and 3 [in the original article]). CIS modules currently on the market reach stable efficiencies of more than 11%. In the laboratory, NREL scientists have achieved cell efficiencies of 19.2% with the semiconductor. Research now focuses on increasing efficiency (Figure 4 [in the original article]), reducing costs, and raising the production yield of CIS panels. Karg predicts that thinfilm technology will eventually halve the present production cost per unit kilowatt peak (kWp), which is the peak power that a solar panel can produce at optimum intensity and sun angle (90°). This implies a cost reduction for a complete system of 35% or more.

In 2000, CdTe solar panels were field-tested on a large scale in the United States. NREL researchers consider CdTe a promising material because of its lower cost of production, which uses techniques that include electrodeposition and high-rate evaporation. Prototype CdTe panels have reached 11% efficiency, and research now focuses on improving efficiency and reducing panel degradation at the electrode contacts. "Studies at Brookhaven National Laboratory strongly suggest that CdTe modules can be safely made in a large-scale manufacturing environment, and that CdTe can be safely disposed of when the modules are eventually retired," Mazer says.

Progress in solar PV research and the development of new applications are guided by national and international collaborations between industry and government, such as those described in the U.S. PV roadmap and carried out by national research teams organized by NREL. Japan and Germany have similar ongoing programs, and leading manufacturers are collaborating with other companies to install solar panels on commercial buildings. For example, a recent agreement between Volkswagen and BP Solar calls for installing solar-energy systems on the roofs of the automaker's dealerships throughout Germany. Each company is investing in several different technologies.

Solar's Big Four

Which PV technologies will dominate future solar-energy markets may depend on the companies developing, manufacturing, and selling them. The four industry leaders are Sharp (Osaka, Japan), BP Solar, Kyocera (Kyoto, Japan), and Shell Solar.

Sharp produces mono- and multicrystalline and amorphous silicon solar cells. The monocrystalline modules have an efficiency of 17.5%, and the multicrystalline cells have 16% efficiency. In 2001, the company shipped 19.2% (75 MWp) of the world's total solar cells. Last July, Sharp opened a new multicrystalline-silicon solar-cell production plant in Nara Prefecture, Japan, and the company's total production capacity now totals 200 MWp.

BP Solar also manufactures nearly 20% of the world's solar-electric panels and systems, using technologies that include polycrystalline solar cells. "We also have our own Saturn technology, which is a highly efficient monocrystalline technology," said a spokesman at BP Solar's U.K. office in Sunbury on Thames, England. The company also sells amorphous silicon thin-film modules. BP developed its proprietary PowerView thin-film silicon laminate partly with funds from NREL. The PV coating converts part of

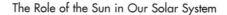

the incoming light into electricity while remaining transparent to the rest of the light. The coating can be used to integrate a solar-energy-generating capacity into building skylights and windowpanes to produce electricity and reduce reliance on utility companies.

Kyocera focuses on off-grid solar systems for private homes in developing countries, communication systems, water pumping, and industry (Figure 5 [in the original article]). It sells its own multicrystalline-silicon systems, and amorphous silicon systems produced by United Solar Systems.

Shell Solar makes mono- and multicrystalline silicon as well as thin-film CIS solar systems. It produced solar panels with a total capacity of about 50 MWp in 2002, and it expects to double its crystalline production capacity by 2004. The company now employs about 900 people in the United States, Canada, Portugal, and Germany after cutting 170 jobs last October.

Although the market growth rate for PV solar panels has declined sharply after four years of annual growth of more than 30%, the growth rate is predicted to be 15 to 20% this year and next. Worldwide production capacity almost doubled last year to 760 MWp, up from 400 MWp in 2001. The main producers of these panels have different business strategies. Shell Solar strongly believes in thin-film alternatives, including its CIS technology. Other companies see crystalline silicon as the dominant technology during the next decade.

Few people doubt solar energy's potential, but many wonder when it will be reached. "In the long term, solar may well play an important role," Karg says. "I personally expect a contribution of 10 to 20% of the global electricity production, mainly in the form of grid-connected systems." However, he does not foresee that happening within the next 20 years.

Reprinted with permission from "Thin Films Seek a Solar Future: Despite Setbacks, the Technology May Yet Shine," Malsch, Ineke, *The Industrial Physicist*, APRIL/MAY 2003, p. 16. © 2003, American Institute of Physics.

"Material Soaks Up the Sun" discusses the use of semiconductors in capturing solar energy for conversion into electricity. A semiconductor is a solid material that has the properties of both an insulator and a conductor. The conducting metal copper, for example, is an excellent conductor of electricity and heat, which is why many cooking pans have a copper bottom. However, copper by itself cannot hold heat for long periods of time. When removed from a hot stove, a copper pan quickly cools. Semiconductors are not as good as metals such as copper in transferring heat and energy, but they provide better insulation, meaning they can more effectively store energy.

Here, Kimberly Patch discusses a new kind of semiconductor made out of doped indium nitride. "Doping" in scientific terms means unnaturally altering a material's atomic construction. Doping can improve the performance of photovoltaic materials, which are energy-transferring

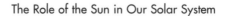

materials of mixed nature or properties, such as semiconductors. Time will tell if indium nitride research leads to the development of better solar cells or if some other product will upstage it in the future. —JV

"Material Soaks Up the Sun"
by Kimberly Patch
Technology Research News, December 11–25, 2002

Capturing solar energy efficiently means finding materials that absorb as many wavelengths of light as possible that the sun sends our way.

Researchers from Lawrence Berkeley National Laboratory, the University of California at Berkeley and Cornell University have discovered that measurements of the semiconductor indium nitride taken two decades ago were wrong. The measurements led researchers to erroneously classify the material as a mediocre photovoltaic.

Instead, the material's band gap falls squarely in the solar spectrum, making it potentially more efficient than currently used photovoltaics. A material's band gap determines which wavelengths of photons can excite electrons in the material to create a flow of electricity.

The reclassified material promises to boost the efficiency of solar cells, allowing for smaller cells or cells of the same size that collect more electricity.

The best photovoltaic material currently available, a combination of gallium arsenide and gallium indium phosphide, has a theoretical efficiency of 32 percent;

indium nitride has the potential to increase that to 50 percent, said Wladek Walukiewicz, a senior staff scientist at Lawrence Berkeley National Laboratory.

The researchers' initial testing also shows that the material withstands high energy particle irradiation without breaking down. This bodes well for satellite and other solar collectors that work in space.

When they made their surprising discovery, the researchers were investigating the mysterious lack of emissions from the material at the conventionally understood band gap of two electron volts. Instead of finding unusual reasons for the lack, they discovered that the material was simply misclassified, and instead has a band gap of 0.7 electron volts. "Once we looked at a lower energy range we could easily see all the features characteristic of [a] direct band-gap semiconductor," said Walukiewicz.

The misclassification took place at a time when samples were prepared by sputtering the material onto a surface. "Such samples contain large amounts of oxygen, as much as 30 percent. So the samples were indium-oxygen-nitride alloys rather than indium nitride," said Walukiewicz.

The more modern samples, in contrast, were grown by molecular beam epitaxy, a process that takes place in a vacuum. "Our material . . . contains undetectable levels of oxygen," he said.

A solar cell works by separating the positive and negative charges—holes and electrons—produced in a semiconductor unit when photons of sunlight hit it. The atomic structure of the semiconductors used in solar

cells determines how many holes and electrons they can generate. The efficiency of a material has to do with microscopic measurements: the band gap, or spacing between the levels electrons occupy in a material, and the wavelengths of solar radiation.

Materials whose electron level spacing matches up with the wavelengths of the photons hitting it can absorb the photons. When a material absorbs a photon, some of the photon's energy causes an electron, which holds a negative charge, to change to a higher-energy position, leaving behind a positively-charged hole.

To separate these charges, the substance must be doped, or mixed with another substance to create separate paths for positive and negative charges. The separate types of semiconductor are n-type, which guides electrons, and p-type, which guides holes. "When an electron-hole pair is produced by a photon . . . the electron is pulled to the n-type side, and the hole is pulled to the p-type side," said Walukiewicz.

Because the charges naturally attract each other, the charge separation creates a change in electric potential, much like rolling a rock up a hill stores energy that can be released by simply starting it on the path downhill. In a solar circuit the potential stored by the separated charges can be released in a current of electricity.

The efficiency of a solar cell depends on how much of the total solar photon flux, or flow, can be converted into charge carriers, said Walukiewicz. "For a solar cell made of one semiconductor with a specific energy-gap, only the photons close to the absorption [range] contribute to the electric current," he said.

To increase the efficiency, however, solar cells can be made by stacking several cells of semiconductors with different band gaps to catch the different light waves. "Although each of the cells will convert the solar energy only from a limited range of photon energies, all of them together can make use of more photons," said Walukiewicz.

Indium nitride is important because its band gap sits squarely in the middle of the light spectrum; this makes it possible to produce gallium-indium-nitride semiconductors that have any gap within the solar spectrum range, and makes it possible to put more semiconductors in tandem, said Walukiewicz. "This is a solar-cell-designer paradise [because] one can maximize . . . performance by optimizing the number of cells and their band gaps," he said.

There's a lot of work to be done before practical solar cells can be made from indium nitride, however. The researchers have not yet made the p-type form of the material. "One of the biggest challenges is to make p-type doped indium nitride," said Walukiewicz. The indications are good, however. It is theoretically easier to make p-type doped indium nitride than to do the same with gallium nitride, which has already been done, he said. Gallium nitride is also a direct band-gap material.

The researchers' next step is to make p-type indium nitride. They are also working to make p-type gallium indium nitride, he said. And they are more thoroughly testing the properties of the two materials under high-energy particle irradiation, he said.

The researchers have only tested a few samples, said Cheng Hsiao Wu, a professor of electrical and computer engineering at the University of Missouri at Rolla. The reasons for the measurements are not yet clear; there could be a mechanism involved other than a different band gap, he said.

Given the benefit of the doubt, however, a 0.7 reading could be useful, but finding such a material is a long way from making a solar cell, said Wu. "For actual solar applications, either the current or the voltage of each solar cell using a particular [part of the] solar spectrum has to be matched," and this is a particularly tricky proposition for a full-spectrum solar cell, he said. If the material works out, however, it "may add another variety of the solar cell we already have for the full visible spectrum range," he said.

It will take three to four years to develop indium-nitride-based solar cell technology, said Walukiewicz.

Walukiewicz's research colleagues were Junqiao Wu and Eugene E. Haller from the University of California at Berkeley and Lawrence Berkeley National Laboratory, W. Shan, Kin Man Yu and Joel W. Ager III from Lawrence Berkeley National Laboratory, and Hai Lu and William J. Schaff from Cornell University. They published the research in the November 15, 2002 issue of Physical Review B. The research was funded by the Department of Energy.

This article reprinted from Technology Research News, www.trnmag.com.

Web Sites

Due to the changing nature of Internet links, the Rosen Publishing Group, Inc., has developed an online list of Web sites related to the subject of this book. This site is updated regularly. Please use this link to access the list:

http://www.rosenlinks.com/cdfa/rsoss

For Further Reading

Birch, Robin. *Sun*. Philadelphia, PA: Chelsea Clubhouse, 2004.

Green, Simon, and Mark Jones. *An Introduction to the Sun and Stars*. Cambridge, England: Cambridge University Press, 2004.

National Research Council. *The Sun to the Earth and Beyond—A Decadal Research Strategy in Solar and Space Physics*. Washington, DC: 2004.

Pasachoff, Jay. *The Complete Idiot's Guide to the Sun.* Indianapolis, IN: Alpha, 2003.

Schielicke, Reinhard. *Reviews in Modern Astronomy: The Sun and Planetary Systems—Paradigms for the Universe.* Vol. 17. New York, NY: John Wiley & Sons, Inc., 2004.

Stix, Michael. *Sun: An Introduction.* New York, NY: Springer-Verlag New York, LLC, 2004.

Upgren, Arthur. *Many Skies: Alternative Histories of the Sun, Moon, Planets and Stars.* Piscataway, NJ: Rutgers University Press, 2005.

Whitehouse, David. *Sun: A Biography.* New York, NY: John Wiley & Sons, Inc., 2005.

Zirker, Jack. *Journey from the Center of the Sun.* Princeton, NJ: Princeton University Press, 2004.

Zirker, Jack. *Sunquakes: Probing the Interior of the Sun.* Baltimore, MD: Johns Hopkins University Press, 2003.

Bibliography

Alerstam, Thomas, Gudmundur A. Gudmundsson, Martin Green, and Aders Hedenstrom. "Migration Along Orthodromic Sun Compass Routes by Arctic Birds." *Science*, Vol. 291, Issue 5502, January 12, 2001, pp. 300–303.

Cowen, Ron. "Stormy Weather: When the Sun's Fury Maxes Out, Earth May Take a Hit." *Science News*, Vol. 159, No. 2, January 13, 2001.

Dyer, Fred C., and Jeffrey A. Dickinson. "Development of Sun Compensation by Honeybees: How Partially Experienced Bees Estimate the Sun's Course." *Proceedings of the National Academy of Sciences*, Vol. 91, pp. 4471–4474.

Frey, H. U., T. D. Phan, S. A. Fuselier, and S. B. Mende. "Continuous Magnetic Reconnection at Earth's Magnetopause." *Nature*, Vol. 426, December 4, 2003.

Gopalswamy, N., A. Lara, S. Yashiro, and R. A. Howard. "Coronal Mass Ejections and Solar Polarity Reversal." *The Astrophysical Journal*, 598: L63–L66, November 20, 2003.

Hasegawa, H., M. Fujimoto, T. D. Phan, H. Rème, A. Balogh, M. W. Dunlop, C. Hashimoto, and R. TanDokoro. "Transport of Solar Wind into Earth's Magnetosphere Through Rolled-Up Kelvin–Helmolz Vortices." *Nature*, Vol. 430. August 12, 2004, pp. 755–758.

Lean, Judith, and David Rind. "Earth's Response to a Variable Sun." *Science*, Vol. 292, Issue 5515, April 13, 2001, pp. 234–236.

Lenz, Dawn. "Understanding and Predicting Space Weather." *The Industrial Physicist*, Vol. 9, Issue 6, December 2003/January 2004, pp. 18–21.

Malloy, Kirk D., Molly A. Holman, David Mitchell, and H. William Detrich III. "Solar UVB-Induced DNA Damage and Photoenzymatic DNA Repair in Antarctic Zooplankton." *PNAS*, Vol. 94, February 1997, pp. 1258–1263.

Malsch, Ineke. "Thin Films Seek a Solar Future: Despite Setbacks, the Technology May Yet Shine." *The Industrial Physicist*, April/May 2003.

Mazza, Carlos A., Jorge Zavala, Ana L. Scopel, and Carlos L. Ballaré. "Perception of Solar UVB Radiation by Phytophagous Insects: Behavioral Responses and Ecosystem Implications." *PNAS*, Vol. 96, Issue 3, February 2, 1999, pp. 980–985.

Metz, Cade. "Solar Flare Could Disrupt Technology." *PC Magazine*, October 24, 2003. Retrieved July 2004 (http://www.pcmag.com/article2/0%2C1759%2C1362967%2C00.asp).

Patch, Kimberly. "Material Soaks Up the Sun." *Technology Research News*, December 11–25, 2002.

Perkins, Sid. "Pinning Down the Sun-Climate Connection: Solar Influence Extends Beyond Warm, Sunny Days." *Science News Online*, Vol. 159, No. 3, January 20, 2001.

Perry, Charles A., and Kenneth J. Hsu. "Geophysical, Archaeological, and Historical Evidence Support a Solar-Output Model for Climate Change." *PNAS*, Vol. 97, No. 23, November 7, 2000, pp. 12433–12438.

Raymond, John C. "Enhanced: Imaging the Sun's Eruptions in Three Dimensions." *Science*, Vol. 305, Issue 5680, July 2, 2004, pp. 49–50.

Weiss, Peter. "Physics Bedrock Cracks, Sun Shines In." *Science News*, Vol. 159, No. 25, June 23, 2001.

Weiss, Peter. "The Sun Also Writhes: Laboratory Solar Physics Sheds First Light on Sol's Seething Sinews." *Science News*, Vol. 155, No. 13, March 27, 1999.

West, Krista. "Storm Spotting: A Step Closer to Forecasting Disruptive Solar Activity." *Scientific American*, March 15, 2004. Retrieved July 2004 (http://www.sciam.com/article.cfm?articleID = 0001F9F6-B662-101EB40D83414B7F0000&sc = I100322).

Xiong, Fusheng S., and Thomas A. Day. "Effect of Solar Ultraviolet-B Radiation during Springtime Ozone Depletion on Photosynthesis and Biomass Production of Antarctic Vascular Plants." *Plant Physiology*, February 2001, Vol. 125, No. 2, pp. 738–751.

Yin, Qing-zhu. "Predicting the Sun's Oxygen Isotope Composition." *Science*, Vol. 305, Issue 5691, September 17, 2004, pp. 1729–1730.

Index

About the Editor

Jennifer Viegas is a news reporter for the Discovery Channel and the Australian Broadcasting Corporation. She has also written for *New Scientist*, Knight-Ridder newspapers, the *Christian Science Monitor*, the *Princeton Review*, and several other publications, as well as a number of books for young adults on a variety of science subjects.

Photo Credits

Front cover (top inset) © Royalty-Free/Corbis; (middle inset) © Digital Vision/Getty Images; (bottom inset) © NASA Jet Propulsion Laboratory (NASA-JPL); (bottom left) © Library of Congress Prints and Photographs Division; (background) Brand X Pictures/Getty Images. Back Cover (top) © Photodisc Green/Getty Images; (bottom) © Digital Vision/Getty Images.

Designer: Geri Fletcher; Editor: Wayne Anderson